D0204778

MENTAL HEALTH IN AMERICA

A Reference Handbook

Selected Titles in ABC-CLIO's

CONTEMPORARY WORLD ISSUES

Series

Adoption, Barbara A. Moe
Chemical and Biological Warfare, Al Mauroni
Childhood Sexual Abuse, Karen L. Kinnear
Conflicts over Natural Resources, Jacqueline Vaughn
Domestic Violence, Margi Laird McCue
Energy Use Worldwide, Jaine L. Moan and Zachary A. Smith
Food Safety, Nina E. Redman
Genetic Engineering, Harry LeVine, III
Gun Control in the United States, Gregg Lee Carter
Human Rights Worldwide, Zehra F. Kabasakal Arat
Illegal Immigration, Michael C. LeMay
Intellectual Property, Aaron Schwabach
Internet and Society, Bernadette H. Schell
Sentencing, Dean John Champion
U.S. Military Service, Cynthia A. Watson
World Population, Geoffrey Gilbert

For a complete list of titles in this series, please visit **www.abc-clio.com**.

Books in the Contemporary World Issues series address vital issues in today's society such as genetic engineering, pollution, and biodiversity. Written by professional writers, scholars, and nonacademic experts, these books are authoritative, clearly written, up-to-date, and objective. They provide a good starting point for research by high school and college students, scholars, and general readers as well as by legislators, businesspeople, activists, and others.

Each book, carefully organized and easy to use, contains an overview of the subject, a detailed chronology, biographical sketches, facts and data and/or documents and other primary-source material, a directory of organizations and agencies, annotated lists of print and nonprint resources, and an index.

Readers of books in the Contemporary World Issues series will find the information they need in order to have a better understanding of the social, political, environmental, and economic issues facing the world today.

MENTAL HEALTH IN AMERICA

A Reference Handbook

Donna R. Kemp

CONTEMPORARY WORLD ISSUES

A B C CLIO
Santa Barbara, California
Denver, Colorado
Oxford, England

Library of Congress Cataloging-in-Publication Data

Kemp, Donna R., 1945–
Mental health in America : a reference handbook / Donna R. Kemp.
p. cm. — (Contemporary world issues)
Includes bibliographical references and index.
ISBN-10: 1-85109-789-9 (hardcover : alk. paper)
ISBN-10: 1-85109-794-5 (ebook)
ISBN-13: 978-1-85109-789-0 (hardcover : alk. paper)
ISBN-13: 978-1-85109-794-4 (ebook)
1. Mental health. I. Title.

RA790.6.K4644 2007
362.2—dc22

2006038850

11 10 09 08 07 1 2 3 4 5 6 7 8 9 10

ISBN-13: 978-1-85109-789-0 (ebook) 978-1-85109-794-4
ISBN-10: 1-85109-789-9 (ebook) 1-85109-794-5

ABC-CLIO, Inc.
130 Cremona Drive, P.O. Box 1911
Santa Barbara, California 93116-1911

This book is also available on the World Wide Web as an ebook. Visit www.abc-clio.com for details.

This book is printed on acid-free paper ♾

Manufactured in the United States of America

Contents

Preface

Mental health has been and still is a problematic policy area. People with mental illness have faced many problems from society throughout the ages. In the past, people with mental illness were often believed to be possessed by demons or the devil and were left in the care of their families or left to wander. They were sometimes mistreated. Eventually, society chose to hospitalize people with mental illness, but their status was reflected in Pennsylvania, where the first mental hospital was placed in the basement of the general hospital. Mental health has continued to be the poor stepchild of the wider health care arena.

The deinstitutionalization movement resulted in the downsizing and closing of many psychiatric hospitals, but treatment in the community has failed to meet the needs of many people with mental illness. In recent years an effort has been made to address the stigmatization of people with mental illness and to provide services to the homeless mentally ill, but a large number of people with mental illness have fallen into the criminal justice system. There remains a need to provide more and better services to people with mental illness in the community.

The 1990s saw the proclaiming of the Decade of the Brain, and much research has been focused on the brain. Many mental disorders are now seen to have a component involving brain chemistry and function. There are those who believe that eventually all mental disorders will be seen as disorders of the brain and will be treated somatically. This could remove much of the stigma related to mental illness.

Worldwide there are differences in how mental illness is viewed and treated. In the countries of the former USSR for instance, large institutions still provide services to psychiatric patients, and in developing countries there is a dearth of any services.

This book looks at the background and history of mental health issues in America and at the policy arena for mental health in the United States, where the states remain the major players in the development and maintenance of the public mental health system. This book examines how mental health has been defined and the most common mental disabilities in the United States and around the world. Treatment, research, prevention, and consumer choice are explored. Issues involving funding, parity, managed care, and integration are examined, and the worldwide perspective is also addressed. A chronology of significant events in mental health is included, and brief biographies of significant people in the mental health field.

Access to the World Wide Web has influenced the facts, statistics, documents, and reports made available through this work as well as legislation and court cases. Organizational sites are now readily available on the Web as well as many print and nonprint resources. The Web brings with it not only access to large numbers of resources, but also the difficulty of determining the value of the material accessed and what are the best sources to use. High school and college students now have many more sources of information available to them, but with this increased availability of sources comes the difficulty of selecting material for understanding the issues involved without becoming overwhelmed by the volume of resources. This book offers a starting point for research by providing an overview of the history of mental health in the United States, a view of some of the current issues in mental health, and a summary of some of the resources for further research.

Finally, I would like to thank the Department of Political Science at California State University, Chico, the department chair, Diana Dwyre, for research time for completing this project, and my editors Dayle Dermatis and Kristine Swift for their assistance and many useful comments during the development and production of this book.

1

Background and History

Theories of how to deal with people with mental illness have shifted over time. Sometimes those with mental illnesses have been ignored and left to manage in the community as best they can. At other times, society has been afraid of them and wanted them shut away. In the best of times, society has tried to help them. The practice of community mental health treatment is traceable to before 400 BC, as Chinese medicine addressed mental function and disorders. The history of mental health is not the same in all countries. Culture, politics, and religion affect how mental health problems are addressed in different countries.

Originally, mental health care occurred in the community and was largely the responsibility of the family, and that has continued to be the case over the centuries in some countries, such as India. Before the fifteenth century in Europe, mental illness was largely ignored, and people with mental illness were neglected. Bizarre behavior was seen as deviant, but not requiring confinement. To the extent that people with mental illness were cared for within the community, that care was financed by individual and religious charity and government aid to the poor. In Ghell, Belgium, to take one example, the community cared for persons with mental illness for over 600 years without ever resorting to institutions.

Institutional care began in a few Arab countries; asylums were established as early as the eighth and ninth centuries to care for people with mental illness. Somewhat later in Europe, but still during the Middle Ages, the community began to seek confinement of people who were different, and some monasteries housed the mentally ill, usually treating them well. In the later Middle

Ages and the following centuries, mentally ill people were sometimes accused of witchcraft; between the fifteenth and eighteenth centuries, it is estimated that at least 100,000 people were executed as witches, many of whom may have suffered from mental illness (Deutsch 1949, 18). In medieval times, the primary methods of prevention and care were ritual and magic. As societies became more urban and families became less able to care for persons with mental illness, eventually, society chose to hospitalize people with mental illness. In France, the so-called Great Confinement in 1665 resulted in the imprisonment of perhaps as much as 1 percent of the population, including people with mental problems, in the general hospital in Paris (Zilboorg and Henry 1941).

America through the Nineteenth Century

During the colonial period, it was at first primarily the family and the parish that provided community care. Persons not cared for within the family were provided for by individual charity. If dependent on public support, they were sometimes housed in private homes at the town's expense, and sometimes even relatives were paid so that the person could remain at home. At this time, religion played an important role in the definition of mental illness; many people perceived those with mental illness as possessed by the devil. Occasionally people with severe mental illness were driven out of communities or killed.

As towns grew, care moved away from private homes to almshouses for the poor and mentally disabled, who fell under the system of poor laws. Colonies passed laws giving town councils the power to care for the "insane" and "feebleminded." In Rhode Island, the colony passed a law in 1742 that allowed town councils to assume responsibility for the care of insane and feebleminded persons and to name guardians for their estates (Nutting 1902). Care, or lack of care, was to continue to be offered through this approach until the nineteenth century. Those who were mentally ill and violent were put in jails, and other persons with mental illness were increasingly placed in workhouses, almshouses, and houses of correction. Conditions in poorhouses were bad, with little in the way of medical care. A distinction was not made between the "insane" and the "feeble-minded" (people with mental retardation). Both were placed under the category of "mentally defective" and were treated the same (Deutsch 1949, 332).

General hospitals were founded in the late eighteenth century, and Pennsylvania Hospital, in the mid-eighteenth century, became the first hospital in the United States to admit mental patients. Their low status can be seen as reflected in the fact that they were treated in the basement of the hospital. The first hospital established only for persons with mental illness opened in Williamsburg, Virginia, in 1776. The establishment of a psychiatric hospital in Pennsylvania and other states soon followed. The establishment of these hospitals was also the beginning of a two-tier public health policy, which in many instances continues today, in which mental health care is frequently segregated from physical health care and is often treated as the poor stepchild, with less funding.

By the late eighteenth century, some physicians began to consider the relationship between "insanity" and natural causes such as disease and stress. Slowly some people also began to recognize mental disability as an economic and social problem (Rochefort 1997), but it was the former attitude that had an immediate effect. One important pioneer who influenced developments in the young United States was Dr. Philippe Pinel, who after the French Revolution took charge of two psychiatric hospitals in Paris and began an approach, called moral treatment, that was based on providing a sympathetic and supportive environment to assist healing. This approach of permissiveness and kindness was followed in the small early mental hospitals in the United States. Many people, however, were still placed in urban almshouses and jails. In response to those conditions, Dorothea Dix began in the mid-nineteenth century a reform movement by advocating the building and expansion of state psychiatric hospitals, so as to allow persons with mental illness to be moved from the poorhouses and jails into an environment in which they could receive care. For financial reasons, the almshouses and jails quickly released the mentally ill to the new state hospitals. Social reformers after the Civil War supported the asylum movement along with their other causes.

The trend of viewing mental illness in natural terms that had begun among physicians in the eighteenth century continued in the nineteenth century, and one contributing factor was a widespread acceptance, by the early nineteenth century, of phrenological theory. Phrenology held that the mind was composed of several distinct faculties that were located in specific areas of the brain. Insanity was believed to be caused by actual physical lesions of the brain. Insanity also had a psychological component, in the sense that it could be caused by environmental and emotional factors (Dain 1964).

In 1828 Horace Mann, an educational reformer, had put forward a philosophy of public welfare that called for making the "insane" wards of the state. This philosophy was widely put into effect, and each state assumed responsibility for the "insane" in that state. States often built their psychiatric hospitals in rural areas, and farms were made a part of the institutions. The farms provided some of the food for the patients, and some of the patients worked on the farms. Moral treatment and compassionate care were the main approach at this time, but with rapid urbanization and increased immigration, the state mental health systems began to be overwhelmed. Many elderly people who in rural areas would have been cared for at home could no longer be cared for when their families moved into the cities. Women as well as men frequently worked away from home, and there was no one to care for the elderly or see to their safety. Many people with brain-based dementias, probably caused by Alzheimer's or small strokes, became patients in mental institutions for the remainder of their lives. The institutions also had many cases of people in the last stages of syphilis. Many of those suffering from mental retardation, epilepsy, and alcohol abuse were also committed to the institutions; in hard economic times, the number of people admitted to the institutions went up.

In the move to moral treatment, the asylum was meant to insulate the mentally ill from the pressures of community life. Moral treatment was the type of psychological treatment provided. It was believed that if environmental factors could cause mental illness, they could also cure it through humane and individualized treatment. A homelike environment was provided with programs of recreation, education, and religion. It is probable that this was a time of the best rates of improvement from the early 1880s to the 1950s (Grob 1973). When the hospitals were small and rural and took on the philosophy of moral treatment, they tried to provide shelter and good care. It was as they became so large and overcrowed that the hospitals became very bad.

Across the country, the various states took up their responsibilities. In 1848, Massachusetts founded the nation's first institution for the mentally retarded, the Massachusetts School for Idiotic Children and Youth. As other states established similar institutions, the number of people with mental retardation in the state mental hospitals decreased.

In California the first recorded mental institution was a converted prison ship grounded in San Francisco Bay. State mental hospitals were developed in 1851 in Sacramento and Stockton. Over the next century, California, like other states, developed a system of farm-based institutions devoted to moral treatment based on concepts of compassionate care. In 1860 at the California Insane Asylum, 40 percent of the patients had been hospitalized for less than a year, while only 0.1 percent were hospitalized for five years or more (Grob 1983, 5).

By 1861 there were forty-eight mental hospitals, with twenty-seven of them being state facilities, and one federal hospital in Washington, D.C. (Caplan 1969, 10). D.J. Rothman (1971) called this growth in state mental institutions "a cult of asylum," which was built on the social reform movement of the antebellum period. This reform movement heightened awareness of the number of people with mental illness and led to a wave of hospital building.

In the second half of the nineteenth century, attitudes changed again and group and treatment practices deteriorated. As more and more people were admitted to the institutions, the focus changed from treatment to custodial care. Commitment laws sent the dangerous and unmanageable to the state hospitals. More patients were alcoholic, chronically disabled, criminally insane, and senile. Treatment practices deteriorated. The institutions became overcrowded, and by the late nineteenth century, the state hospitals were places of last resort, with mostly long-term chronic patients. Better treatment was found in small private psychiatric hospitals for those who could afford the care, but most people had to go to the large psychiatric hospitals, where conditions remained bad for many years. An analysis of commitment proceedings in San Francisco in the early twentieth century revealed 57 percent were begun by relatives, 21 percent by physicians, and 8 percent by the police (Fox 1978, 15). The composition of the patients in the mental hospitals changed. Massive immigration to the United States led to a growing proportion of foreign-born and poor in the state hospitals. Most psychiatrists, community leaders, and public officials were native-born and generally well off, thus apt to be prejudiced against those who were neither (Rochefort 1993). Between 1860 and 1890, seventy-seven new state mental institutions opened (McGovern 1985, 25).

As the nineteenth century drew to a close, a new idea, promoted by what is known as the Eugenics Movement, took hold. This movement held that "insanity" and "feeblemindedness" could be inherited. Professional conferences, humanitarian groups, and state legislatures increasingly identified insanity as a special problem of the poor. Feeblemindedness was identified as a cause of social maladies like poverty, crime, and sexual misbehavior (Grob 1966). Insane persons were increasingly seen as possibly violent and incurable, and as a threat to the community (Caplan 1969). These beliefs led to numerous state laws restricting the lives of people with mental disabilities, including involuntary sterilization laws and restrictive marriage laws. Connecticut in 1896 passed the first restrictive marriage law, and Indiana in 1907 passed the first sterilization law. Other states quickly followed suit. As a result, 18,552 mentally ill persons in state hospitals were surgically sterilized between 1907 and 1940. More than half of these sterilizations were performed in California (Grob 1983, 24). Instead of the previously used psychological therapy, mental hospitals turned to the use of mechanical restraints, drugs, and surgery. Psychiatrists spent more time diagnosing large numbers of patients rather than delivering individualized care. State legislatures did not increase budgets to meet the needs of the growing hospitals. Physical plants became overcrowded and deteriorated. Salaries were not adequate to attract good personnel. Superintendents no longer saw patients but spent their time on administrative tasks, and their influence declined as they became subordinate to new state boards of charity, which were focused on efficiency (McGovern 1985).

The late nineteenth century saw other important developments, which only affected mental health care later. Some psychiatrists were asking questions about the relationship between mental disorders and environment, and Sigmund Freud and others were exploring the idea of an "unconscious" that contained repressed psychic material. From this work emerged psychoanalysis and other therapies.

The Twentieth Century

Gwen was the last child born to a family of seven in the 1920s. As a child she fantasized a lot and did not do well in school. In her teens she married, but the marriage was annulled. In her early twenties she married again, and her husband told her family that she was "crazy," that she flew into rages and threw things. That marriage ended in divorce. Also in her early twenties, she worked as a dispatcher for a taxi company. One day she thought two women on the street were talking about her, and she attacked them. The police took her into custody.

Gwen's family did not know what to do with her. The family doctor advised her mother to commit her to the psychiatric hospital. Her mother and the doctor signed the papers and she was involuntarily admitted to a hospital that was fifty miles from her home. At the hospital, she was diagnosed with schizophrenia. She was given electroshock therapy, but it was not effective for her. At the time that she was there, the newest treatment was lobotomy. They had her mother sign the papers giving permission to perform a lobotomy. She was not the same after the lobotomy, and when they sent her home, her mother was afraid of her.

Her mother recommitted her to the mental hospital, and throughout the 1950s and early 1960s she remained at the hospital. Finally during the deinstitutionalization movement, she was moved to a care home for the mentally ill in a small community. She remained there the rest of her life. The lobotomy left her with no chance for recovery and return to a normal life.

The twentieth century saw many hopeful new developments in the care and treatment of the mentally ill, but in general the promise of these developments was never fully realized, whereas treatments such as lobotomy, which promised to provide a quick and inexpensive fix for mental illness, were widely practiced. How devastating this kind of treatment can be is suggested by the story with which the section begins.

In the early 1900s, the reform movement called progressivism became an important force in the United States. Among the areas of interest of progressivism was poverty; progressives held the belief that dependency came not just from individual problems, but from social and economic forces. At the same time, neurologist and psychiatrist Adolph Meyer developed a new definition of insanity as a disease of social functioning that included difficulties in interpersonal relations (Rothman 1980).

The names of the departments responsible for the mentally ill began to change. As late as 1913, jurisdiction over the mentally ill in California rested with a Commission on Lunacy, but in 1921, California established a Department of Institutions. This department saw as one solution to the growing numbers of mentally ill the deportation of immigrants found to have mental illness. The California Department of Institutions reported deporting 4,160 insane persons and 2,987 delinquents between 1923 and 1929, for a savings of $8.5 million. In 1939, the 262 mentally ill persons deported made up 1.2 percent of the total institutional population (California Department of Institutions 1939, 27).

The first half of the twentieth century saw some promising new treatment developments and attempts to establish community-based systems of services. New therapies included music therapy and photochromatic therapy, or the exposure of patients to sunlight filtered through colored glass. The custodial institution remained the main site of care, but some institutions developed cottage systems that placed more able patients in small, more homelike structures on the hospital grounds. Another new development was the creation of family care programs for boarding outpatients. In 1909 Clifford Beers founded the National Committee for Mental Hygiene to encourage citizen involvement, prevention of hospitalization, and aftercare for those who left the hospitals. All these approaches, however, only served a small part of the population.

A few psychopathic hospitals for short-term care and a few outpatient clinics for ambulatory care were developed in the early 1900s. By the 1920s and 1930s, there was some care for children in the community by a few child guidance community clinics. Although the Division of Mental Hygiene was created in 1930 in the U.S. Public Health Service, it did not address institutional or community mental health care in general but only narcotics addiction.

Although the concept of community placement was discussed in California in the 1920s, by 1938 there were still 22,000 patients in California mental hospitals. At that time, the new head of the Department of Mental Health, Dr. Aaron Rosanoff, convinced the governor and legislature to set up an extramural program to initiate releases, a program financed through lowered state hospital census. Over a two-year period, 2,000 patients were placed in the community, and the U.S. Social Security Board was convinced to provide public assistance to elderly patients "on leave" from the hospitals (Barter 1983, 1960). With the advent of World War II, however, community placement declined.

During the Great Depression of the 1930s and World War II, few resources were available to psychiatric institutions, but they continued to grow anyway. Some hospitals had as many as 10,000 to 15,000 patients. From 1930 to 1940, the number of people in state mental hospitals increased five times faster than the general population to a total of 445,000 (Rothman 1980, 374). With an increasingly urbanized population and with people living longer, a fifth of some hospital admissions were people sixty or over (Grob 1983, 43).

Four new therapies arose in the 1930s: insulin coma therapy, metrazol-shock treatment, electroshock therapy, and lobotomy. These treatments were given to thousands of patients, in many cases with devastating results; nevertheless, they were widely used until the appearance of antipsychotic medications in the 1950s.

Shock treatments included electric shock and high doses of insulin and Metrazol (pentylenetetrazol), which were used to induce comas and were intended to jolt depressed or hallucinating patients out of their mental illness. Some patients were helped, but many suffered bone fractures, brain damage, and chemical poisoning.

Between 1936 and 1960, an estimated 50,000 lobotomies were performed in the United States. A Portuguese neurologist, Antonio Egas Moniz, developed the procedure. In 1936, he reported that he had cut into the brain lobes of psychiatric patients, including schizophrenics, to try to cure them. His idea was extreme but not original. Psychosurgery can trace its roots to prehistoric attempts at trepanning by the Incas and other tribes, who believed that evil spirits could be evicted from the mind and soul by creating openings in the brain to let them out. Early Romans also observed that a sword wound to the head occasionally cured insanity. Modern psychosurgery began in the 1880s, when Gottlieb Burckhard, a Swiss psychiatrist with his own mental hospital, removed pieces of cortex to calm agitated patients. An Estonian surgeon, Lodivicus Puusepp, abandoned the idea after three similar operations in 1910. There was no way to see inside the brain, so all of these operations were blind. In 1920, Johns Hopkins neurosurgeon Walter Dandy developed x-ray techniques that allowed outlining the main cavities of the brain. In 1932 he removed a brain tumor from the frontal lobe of a young stockbroker, which resulted in significant personality changes. In 1933, Carlyle Jacobsen, a protégé of Fulton, and John Fulton, the chairman of Yale University's Department of Physiology, conducted lobotomies on two chimpanzees that led them to conclude that frontal lobe loss caused erratic and unpredictable changes in learning and behavior. Egas Moniz, a Portuguese neurologist, attended the Second International Neurological Congress in 1935 and heard about the chimpanzee experiments. Moniz and his neurosurgeon, Almeida Lima, performed their first lobotomy on a woman patient in 1936. In 1937, Moniz reported on twenty psychotics and neurotics who received a cut across the whole top of the cortex in the front of the brain, claiming cure for a third of his patients, improvement in a third, and no change in a third (Rodgers 1992). In 1949 Egas Moniz won the Nobel Prize for his work on lobotomies.

The most ardent supporter of lobotomies in the United States was Walter Freeman, a professor of neurology at George Washington University who was not a certified surgeon. He was the leading lobotomist in the United States. His colleague James Watts, a neurosurgeon, assisted him. Freeman and Watts performed the first U.S. lobotomy in September 1937. They published the first textbook on the subject in 1942, *Psychosurgery in the Treatment of Mental Disorders and Intractable Pain.* They experimented on their patients, changing the procedure as they went. By 1946 Freeman had become dissatisfied with prefrontal lobotomies. The standard procedure for prefrontal lobotomy was to cut open the skull or drill holes so that brain tissue could be scooped out with a blunt metal instrument or killed with alcohol injections. The method required a hospital and surgical team. After practicing on cadaver brains, Freeman found he could use an ice pick to quickly and easily perform lobotomies without a surgical team. He did the first of thousands of so-called transorbital lobotomies on a woman named Ellen (Rodgers 1992). The surgery consisted of using two to five successive electroconvulsive shocks to maintain a state of coma for about five minutes. The ice pick, called a leukotome in scientific reports, was thrust up between the eyeball and eyelid through the roof of the eye socket, or bony orbit, into the frontal lobe of the brain. The ice pick was then manipulated to damage the brain tissue.

Freeman developed a huge private practice in Washington, D.C. He then traveled around the country performing the procedure. During the summers he took his children camping across the country and on the way stopped at psychiatric hospitals to teach psychiatrists how to perform the procedure. He performed lobotomies in fifty-five hospitals in twenty-three states, as well as Canada, the Caribbean, and South America (Rodgers 1992, 13). The procedure was performed on patients with a wide variety of conditions, including depression, anxiety, phobias, paranoia, schizophrenia, psychosis, neurosis, hyperactivity, frigidity, homosexuality, delusions, and violence. The procedure was able to spread so rapidly because so many people with mental illness were confined in psychiatric institutions under appalling conditions. Psychiatrists, surgeons, and those who ran the institutions jumped at the chance to do something that might lead to the alleviation of mental illness and the release of patients back into society. The number of first admissions to psychiatric hospitals grew at the average rate of 80 percent a year during the 1930s. The number of patients in state hospitals was 432,000 in 1936 and 445,000 in 1940. Hospitalization for mood disorders such as depression and anxiety provided the greatest increase, but there were also significant increases in schizophrenia. The prognosis for schizophrenia was considered nearly hopeless, and patients spent years, even decades in institutions, receiving little if any therapy and being kept in a predominantly custodial environment (Shutts 1982, 7).

A. Earl Walker, a former chief of neurosurgery at the University of Maryland School of Medicine, tried in 1944 to evaluate the results of lobotomies in hospitals. He found widely varying success rates and concluded that success or failure with a patient was probably related to the severity of illness of the patient and not to differences in surgeries. There were no rigorous standards of diagnosis and evaluation, and so no way to judge improvements attributed to psychosurgery. He also gathered negative reports: in some cases, the personality was flattened and altered, and there were reports of bleeding, suicide, deterioration of IQ, inertia, sexual aberrations, and aggressive behavior (Rodgers 1992). In a speech at the 1946 American Psychiatric Association, psychologist Ward Halstead noted the lack of scientific methodology in the evaluation of patients. Between 1949 and 1951, the National Institute of Mental Health published critiques that pointed out the deficiencies of follow-up studies, including lack of controls, poor quality of evaluations, and imprecise criteria for outcomes. When research began to be done in the 1950s and 1960s on patients who had previously received lobotomies, the resulting studies were notably negative. Eventually the lack of rigorous diagnosis, evaluation, and follow-up, the brutality of the ice-pick surgeries, the growing consumer and patients' rights movements, and the advent of neuroleptic, or antipsychotic, drugs led to the demise of lobotomies. In 1971 Freeman finally published a long term follow-up of 707 schizophrenics operated on four to thirty years earlier. Seventy-three percent were in psychiatric hospitals or in a state of dependency at home (Rodgers 1992). States began passing laws outlawing the practice of psychosurgery, including Oregon in 1973 and California in 1976.

By 1940 public psychiatric hospitals cared for 98 percent of all institutionalized patients. There were more than 400,000 patients in state psychiatric facilities and an additional 50,000 psychiatric patients in veterans', city, and county institutions (Grob 1987, 412). Fifty percent of the hospital beds in the country were occupied by the mentally ill (Rochefort 1993, 31–32). A substantial percentage of state budgets was spent on institutional care.

Although with the advent of World War II reliance on community placement at first declined, the emphasis on short-term and community-based care returned, encouraged by the military's interest in psychological testing and the high numbers of men they rejected from the draft for mental health reasons. They were also interested in returning the large number of soldiers with combat-related stress disorders to combat or the community, and so they were interested in quick and effective treatment.

Another factor in promoting change after World War II was a series of exposés on the scandalous conditions in state mental hospitals; as a result, community clinics that had been largely serving children and adolescents began to serve adults as well. But there were many factors that led to changes in mental health policy after 1945, too many to discuss individually; an extensive analysis (Grob 1987) identified the following factors:

- The change in the composition of the patient population, 1890 to the 1940s, from a population of

acute cases largely institutionalized for less than a year to a population consisting mostly of long-term chronic patients, with many aged patients, behavioral problems, and dependent persons in custodial care
- A growing division between psychiatrists, who began to see mental illness as preventable through early treatment in community facilities, and the traditional mental hospitals
- Changing definitions of mental health, with a rejection of the traditional distinction between mental health and mental abnormality
- The growing importance of scientists and academics in formulating and implementing policy.
- The New Deal expansion of the role of the national government
- The decline of eugenics, with its view that heredity causes mental illness, the growing emphasis on the importance of environment, and increased interest in social activism
- The development of activism in the American Psychiatric Association (APA)
- The growth of a coalition of medical and social activists who worked to get the federal government involved in health policy
- A series of exposés on the horrors of the state psychiatric hospitals
- The growing promotion of community care and treatment by the Council of State Governments, the Milbank Fund, and others
- Experimentation with new mental health policies in New York and California
- Establishment of a biomedical lobby

By the late 1940s, more than two-thirds of the members of the APA were still practicing in public institutions (Grob 1987). It was at this point that the federal government began to take some action in regard to mental health. Surgeon General Thomas Parran requested Robert Felix in 1944 to prepare a program enabling the federal government to respond to the problem of widespread mental disabilities found during World War II (Foley 1975). Felix worked for the Mental Hygiene Division of the Public Health

Service. Along with those who had served as the chief psychiatrists of each branch of the military service during the war, he wrote a national mental health plan. The National Mental Health Act of 1946 brought the federal government into mental health policy in a significant way. The act created new federal grants in the areas of diagnosis and care, research into the etiology (causes) of mental illness, professional training, and development of community clinics as pilots and demonstrations. It also funded demonstration studies on prevention, diagnosis, and treatment. The act mandated the establishment of a new National Institute of Mental Health (NIMH) within the Public Health Service in 1946 to encourage research on mental health, and it created the National Advisory Mental Health Council. The role of the federal government in mental health was now greatly expanded, and the federal government was now promoting development of community care.

The states moved toward reform when in 1949 the Governors' Conference released a report detailing the many problems in public psychiatric hospitals, including obsolete commitment procedures; shortages of staff and poorly trained staff; large elderly populations; inadequate equipment, space, and therapeutic programs; lack of effective state agency responsibility for supervision and coordination; irrational division of responsibility between state and local jurisdictions; fiscal arrangements damaging to residents; and lack of resources for research. In 1954 a special Governors' Conference on Mental Health adopted a program calling for expansion of community services, treatment, rehabilitation, and aftercare.

During the 1950s the states pursued both institutional care and expansion of community services. In the mid-1950s, major deinstitutionalization of hospitals began. The introduction of psychotropic medicines to reduce and control psychiatric symptoms created optimism that some mental illnesses could be cured and others could be modified enough to allow persons with mental illness to function in the community. The concept of a therapeutic community arose, with the hope that cures could come through a supportive community, a return in many ways to the approach of the early small psychiatric hospitals and their moral treatment. The move to community care was also encouraged because institutional care was becoming increasingly expensive, and the states were anxious to transfer costs to the federal government's welfare system.

The discovery and use of psychotropic drugs in the 1950s had a profound impact on the treatment of the mentally ill. Tranquilizing drugs were widely used in the state institutions and played a major role in deinstitutionalization. Early discharge programs became common, and the inpatient census of public psychiatric hospitals continued to steadily decline. The unofficial onset of deinstitutionalization was 1955, which was followed by consistent annual census declines at the psychiatric hospitals (Rochefort 1993, 218–219). More and more mental health professionals began to advocate community care. The new drugs raised the hopes of people with mental illness, clinicians, and families. They made psychiatrists more acceptable to other physicians (Roberts 1967). Because of the apparent success of the drugs, more emphasis was placed on a biochemical view of mental illness. At the same time, studies were showing the importance of socioeconomic factors in the origins and treatment of mental illness.

In 1955 the Mental Health Study Act was passed, leading to the establishment of the Joint Commission on Mental Illness and Health, which prepared a survey and made recommendations for a national program to improve methods and facilities for the diagnosis, treatment, and care of the mentally ill and to promote mental health. The commission recommended the establishment of community mental health centers and smaller mental hospitals. Many progressive recommendations were made in the eight volumes published by the commission between 1958 and 1962. These included using socialization, relearning, group living, and rehabilitation in the treatment of mental illness. The focus was on a social treatment model. The commission also advocated prevention, and they laid the groundwork for the Community Mental Health Centers Act of 1963, the legislation that was the beginning of the community mental health revolution, often seen as the third great revolution in the field of psychiatry in the twentieth century. The first was the emergence of psychoanalysis in the 1910s and 1920s, and the second the introduction of psychotropic drugs in the 1950s (Rubins 1971).

An antipsychiatry movement arose, based largely on the writings of Ronald D. Laing and Thomas Szasz. Laing's first book, *The Divided Self*, was written in the late 1950s and published in 1960. The movement went so far as to declare that mental illness did not exist but was simply a form of social control used to handle those who were different.

One of the most hopeful developments that came out of the new ideas of this period was the passage of the Mental Retardation Facilities and Community Mental Health Centers (CMHC) Construction Act of 1963.

In 1960 John F. Kennedy was elected president. His election raised hopes for an expanded federal role, because he was favorably disposed to government action to promote mental health. He had a sister suffering from mental retardation and mental illness, and the 1960 Democratic platform had affirmed federal support for community mental health. Key White House assistants and officials in the Bureau of the Budget were supportive of legislation. President Kennedy appointed two special working groups to study mental health and mental retardation and formulate legislative proposals, and their recommendations endorsed a policy of deinstitutionalization. Kennedy used the power of his office to advocate a community-based program that largely bypassed the states. On February 5, 1963, the president outlined the final legislative proposal before Congress (Kennedy 1964).

The proposed legislation was a compromise between the public health and medical models and argued for an expansion of both community services and individual services. It also took an aggressive anti-institutional approach. Kennedy's speech emphasized that there were about 600,000 persons in state and private institutions and that the public spent $1.8 billion annually to care for them. The president decried the custodial nature of the institutions and their poor conditions. He argued that new therapeutic techniques and especially psychotropic drugs allowed for new directions in community care and stressed the importance of prevention. The goal of the legislation was to strengthen existing resources as well as fund new services. Included in the proposal were grants for training of professional personnel, expansion of research, and upgrading of institutional care. Federal matching grants to the states were to enable construction of Comprehensive Community Mental Health Centers. Planning grants and short-term project grants for initial staffing were also proposed for the centers. The Comprehensive Community Mental Health Centers were to include emergency psychiatric care, outpatient and inpatient services, and foster-home care. A goal was set to reduce custodial care in state mental hospitals by 50 percent in one to two decades.

The program was introduced to the Senate as S. 755 in early February and to the House as HR 3688. Hearings were held in the

Senate Committee on Labor and Public Welfare and the House Committee on Interstate and Foreign Commerce. Many witnesses were heard from the administration, state governments, professional associations, and voluntary associations. Dr. Jack Ewalt from the American Psychiatric Association testified that the new mental health program probably would not produce budgetary savings because of the large numbers of people with mental illness in the community who were not being treated. A joint statement of the American Psychiatric Association and the American Hospital Association called mental illness the number one health problem in the country, afflicting as many as 10 percent of the population in urban areas (U.S. Congress, Senate 1963).

David A. Rochefort (1993) identified several themes that occurred during the hearings. The first theme was the large numbers of people in the population with mental illness, exposing the myth that only a small number of deviant people were mentally ill. A second theme was the bad conditions in the state hospitals. A third theme involved comparing mental and physical illnesses, urging that mental illness be treated the same as physical illness in the community. There was emphasis on mental illness as an illness, which should not lead to the condemnation of individuals. The fourth theme stressed the role of social influences in causing and potentially helping to alleviate mental illness. Several speakers emphasized that separation and isolation of persons with mental illness intensified the illness and turned people into chronic patients.

The American Medical Association opposed the provisions to fund staffing in the centers as part of their opposition to socialized medicine. The staffing provisions were subsequently removed from the legislation. Ultimately, both houses of Congress passed the legislation and appropriated $150 million for implementation, and the president signed the legislation into law on October 31, 1963. The CMHC Act authorized three years of federal grant support for construction of local community mental health facilities. From 1963 to 1980, the legislation was repeatedly amended to extend its life. The first amendment in 1965 added funding for the employment of professional and technical personnel within the community mental health centers. In 1968 the Alcoholic and Narcotic Addiction Rehabilitation Amendments provided funding for building and staffing special facilities for the treatment of alcoholism and addiction. In 1970 special funds were set up for children's mental health services and for education and consultation.

During the remainder of the 1960s the civil rights movement and public interest law strengthened mental health policy and encouraged community mental health treatment and the decline of the role of psychiatric hospitals. The policy of community treatment was reinforced by changes in hospital policies, including open door policies, informal admissions, increased use of psychotropic drugs, and preparation of patients for early release.

The belief that the community should be involved in caring for the mentally ill became ever more widely accepted in the 1960s. Two assumptions underlay this belief: "Close, long-term human relationships, particularly those within small groups, are valuable and to be fostered," and "the strength which comes from humans banding together in social groups is to be prized and utilized" (Zusman 1975, 25). Mental health advocates used these concepts to attack the hospital system.

The passage in 1965 of Medicaid and Medicare stimulated the growth of skilled nursing homes and, later, intermediate-care facilities. It opened the door for what became known as the transinstitutionalization movement, which resulted in the movement of large numbers of persons with mental illness and mental retardation out of institutions and into nursing homes and other community institutional settings. By placing these people in institutions with federal funding, or by deinstitutionalizing them and signing them up for federal assistance programs, state officials could shift costs to the federal government. Inpatient and outpatient psychiatric services expanded in acute-care general hospitals because of increased third-party reimbursement by the government and by insurance companies.

Under the new system, community mental health centers were supposed to replace psychiatric hospitals as the primary providers of care, but there were problems with defining their roles and the populations they should serve. There were even greater problems with funding them. From 1965 to 1969, only $260 million was authorized for community mental health centers. During the Nixon administration, mental health center funds were impounded, and from 1970 to 1973, only $50.3 million was appropriated. The National Council of Community Mental Health Centers brought suit against the administration. The U.S. District Court for the District of Columbia found the impoundment illegal.

By 1980 approximately $2.5 billion had been allocated under the act, funding 796 centers; in 1980, 740 were operating. The centers covered a little more than half of the 1,500 catchment areas in the country accounting for about 115 million persons. About 3 million patients received services annually (Rochefort 1993, 63). The system was criticized on the grounds that not enough community mental health centers had been developed, and that those established focused too much on seeing people with less serious mental health problems and not enough on providing services to the chronically and seriously mentally ill. Although much change occurred, no coherent overall policy was developed.

Deinstitutionalization

With or without a coherent policy for treating people in the community, the movement toward deinstitutionalization continued. After 1965 psychiatric hospital populations declined rapidly, and in the late 1960s and early 1970s, numerous institutions throughout the United States closed their doors or sharply reduced the number of patients they served. Resident populations in state and county mental hospitals declined from 512,501 in 1950 to 101,402 in 1989 (Rochefort 1993).

So far, we have mostly looked at the federal actions that contributed toward deinstitutionalization, but the states played a role too, California in particular. The state was among the first to deliberately legislate the move of mental health patients from state hospitals to communities. In 1957, the Community Mental Health Services Act (Short-Doyle) provided 50 percent funding for locally administered mental health programs. In 1963 the state share was increased to 75 percent. During the 1960s, the focus of Governor Ronald Reagan's administration became deinstitutionalization and closing large state hospitals. Between 1959 and 1969, the state hospital census declined from 38,000 to 16,500. In 1969, the Lanterman-Petris-Short (LPS) Act was passed, increasing the state funding share to 90 percent for local programs and encouraging the trend toward community-based care and treatment. The act was also hailed as a model for restoring the civil liberties of persons alleged to be mentally ill (Bardach 1977, 56). The legislation was considered revolutionary for setting up strict criteria and legal procedures for involuntary hospitalization. The law established standards for involuntary commitment involving a first-stage 72-hour emergency hold, followed by 14-day and 90-day certifications, and then conservatorships (Warren 1982).

By 1973, 80 percent of the state hospital population had been moved out of the hospitals, and a committee of the California Senate studied the possibility of closing them (*Final Report* 1974, 16). However, Senate hearings indicated that community mental health programs were not effective in caring for the chronically ill patients discharged from state hospitals, and a decision was made to keep state hospitals (*Final Report* 1974). By 1980, the hospital census had declined almost 90 percent to under 4,000 people. In California the number of residents in state hospitals declined from approximately 37,000 in 1958 to 1,800 in 1990 (California Council of Community Mental Health Agencies n.d., 7).

California was not alone in recognizing that deinstitutionalization was not a complete answer. By the late 1970s, problems resulting from deinstitutionalization were becoming more and more apparent. Residents were released from institutions, but community-based facilities designed to provide treatment were not established in sufficient numbers. This opened the door for private profit makers who "bid" for the former residents and their state and federal benefits and in many cases provided only minimal, low-quality care. Many patients released from psychiatric hospitals quickly fell between the cracks and became part of the homeless population or ended up in jail. Without adequate services and medications, many of these people self-medicated by using illegal drugs and alcohol, adding to their problems.

California was not a major participant in the development of federally funded community mental health centers, but it did attempt to handle the problems arising from deinstitutionalization—unsuccessfully. Even by 1976, California received only $17 million in federal community mental health funds, and only 14 of 58 counties had established Community Mental Health Centers (CMHCs). California developed most of its community mental health services with state funding, and in 1976 state general fund mental health expenditures were $428 million. With the closing of some institutions and with limited beds available in the remaining hospitals, there developed a need for 24-hour acute treatment facilities. Local mental health programs had begun to develop such facilities in the 1970s. In 1978, Senate Bill 1496 (Chapter 1234, California Health and Safety Code) established licensure and standards for psychiatric health facilities (PHFs) in nonhospital settings for acute psychiatric patients, PHF programs mainly served short-term acute public patients, with 71 percent of PHF admissions involuntary (Moltzen, Gurevitz, Rappaport, et al. 1986, 1132). In 1979 AB 3052 (known as the Bates Bill) funded a few programs around the state at levels near those needed for comprehensive community mental health, but development of the community mental health program had effectively ended by 1975. Follow-up legislation was never passed. In 1979, a legislative task force did create what was known as the California Model, which defined, on a population basis, program elements needed for an adequate unified mental health system (*Report of the Mental Health Legislative Work Group* 1981). In 1982, however, it was estimated that $362 million would have to be added to the existing $600 million mental health budget to meet the California Model standard, and the California Model was never funded.

As deinstitutionalization increased, people with mental health problems were seen more frequently in outpatient clinics rather than in institutions. By the 1980s in the United States there were many more outpatient episodes than inpatient episodes (mental health visits), but twice as much money was still being spent on inpatient care as outpatient care. The total number of inpatient episodes was almost 3 million, and the most frequent inpatient site for mental hospitalization was the general hospital. The most episodes occurred in general hospitals without separate psychiatric units. Over a fifteen-year period, the absolute number of inpatient episodes increased about 60 percent or a million cases. The increase in rate was due to patients being treated in general hospitals without psychiatric units where their stay was the shortest length of stay of any site (Kiesler and Sibulkin 1987, 272). Patients were being actively treated in general hospitals without psychiatric units (Friedman 1985). About 25 percent of all inpatient days in all hospitals were for diagnosed mental disorders, and that percentage had decreased from over 50 percent thirty-five years before. However, the readmission rate to hospitals had not changed. There was no evidence of a revolving door of patients being released and readmitted (Kiesler and Sibulkin 1987, 272).

The failure to address the problems created by deinstitutionalization and the continued overfunding of inpatient care and underfunding of outpatient care were not caused by any lack of evidence that community care is more effective. At least two important studies pulled together evidence that services are more effective in the community (Bandura 1978; Mosher 1983). C.A.

Kiesler (1982), after a review of literature comparing mental hospitalization with alternatives to hospitalization, concluded, "in no case were the outcomes of hospitalization more positive than alternative treatment. Typically alternative care was more effective regarding such outcome variables as psychiatric evaluation, probability of subsequent employment, independent living arrangement, and staying in school, as well as being decidedly less expensive" (Kiesler 1982, 349).

Two types of studies have examined where people have gone when released from institutions: hospital reports and statistics on where patients are discharged and community studies describing the status of former hospital patients. Commonly used categories of living arrangements are alone, with family and relatives, and in supervised living, including board and care homes and halfway houses. Early studies in 1969 showed 70 percent of discharged people returning to families, with about half to spouses and half to other relatives (Group for the Advancement of Psychiatry 1978). One study found 32 percent living with family members, 43 percent living with other clients, and 15 percent living alone (Goldstrom and Manderschied 1982, 33) while another study found 25 percent living alone, 23 percent with parents, 18 percent with spouses, 10 percent with other relatives, 14 percent in board and care homes, and the rest with friends or in other arrangements (Lamb and Goertzel 1977, 680).

In the 1980s, mental health programs tried to respond to people who had not been successfully deinstitutionalized and to the new population of persons with mental illness who had never been in the state institutions but were appearing on the streets and in the jails. Programs attempted to do outreach and rehabilitation and to provide stabilization. Some of these programs will be discussed in the next subsection.

Institutional Reform and Community Mental Health

In 1971, Title XIX of the Social Security Act (Medicaid) was amended to require institutional reform and the meeting of accreditation standards by facilities in order to receive federal funding. To qualify for Medicaid, an individual was required to be disabled, as determined under the Supplemental Security Income (SSI) program, and to meet state income and resource standards.

Persons with mental illness could qualify for services in an Intermediate Care Facility/Psychiatric Rehabilitation (ICF/PR).

In 1975 Congress required community mental health centers to provide aftercare services and mandated linkages between community mental health centers and state hospitals to ensure continuity of care as a person left a psychiatric hospital and went into the community. Also in 1975, the passage of PL 94-142 mandated that a free and appropriate education be available to all handicapped children from three to twenty-one years of age, including those residing in public institutions. Those handicaps could include emotional or behavioral problems if they had an impact upon a child's ability to learn. Full implementation did not occur until 1980, but the act immediately encouraged focus on early intervention and educational settings.

In 1977 President Jimmy Carter, using an executive order, set up the President's Commission on Mental Health, with his wife Rosalyn as honorary chairperson. Its mandate was to conduct a one-year examination of the nation's mental health system and to make recommendations for legislation. In 1979 parents of persons with mental illness founded the National Alliance for the Mentally Ill (NAMI). In the same year the President's Commission on Mental Health recommended several measures: local communities should be allowed the flexibility to provide only the services needed; medical and mental health services should be integrated through more consultation between agencies and caseworkers; new financial resources should be made available for mental health services; and the needs of underserved groups, including children, the elderly, and people in rural areas, should be met. Many of the recommendations were incorporated in the Mental Health Systems Act (PL 96-398), which became law in October 1980. The legislation provided federal funding for ongoing support and development of community mental health programs. New grant programs focused on the chronically mentally ill, children and youth, the elderly, and rural and minority populations. The states were given an augmented role in review and management of federal grant programs. Provisions were set out for protection of mental patients' rights, and greater coordination was specified between health and mental health planning.

The new legislation required statewide plans for mental health systems, but there was no real opportunity for the new act to be implemented. In 1981 the Omnibus Budget Reconciliation Act of the Reagan administration rescinded much of the legislation and revamped the rest. Twenty-one categorical health grants were lumped into four block grants, including one for mental health and substance abuse that defined twelve services. The mental health and substance abuse categorical grants, which had gone directly to local agencies, were consolidated and were distributed by the state governments. Expenditures for the alcohol, drug abuse, and mental health administration went from $1,105,477 billion in 1979 to $761,007 million in 1982 (Hudson 1993). The fiscal erosion of the 1970s and the 1980s thus took a heavy toll on state and local mental health programs.

In 1990 only one in five of those with mental illness received treatment (Castro 1993, 59). Nevertheless, the National Institute of Mental Health estimated the cost of treating mental illness at $148 billion, which included $67 billion for direct treatment (10 percent of all U.S. health spending) and $81 billion for indirect costs such as social welfare and disability payments, costs of family caregivers, and morbidity and mortality connected to mental disorders (Castro 1993, 60).

In the Department of Health and Human Services (DHHS), from which federal funding still largely comes, mental health programs were organized in 1992 into the Substance Abuse and Mental Health Services Administration (SAMHSA), consisting of the Center for Mental Health Services, the Center for Substance Abuse Prevention, and the Center for Treatment Improvement. The institutes on mental health, drug abuse, and alcohol and alcohol abuse were shifted to the National Institutes of Health and began to focus only on promotion of research in mental health and substance abuse. The Department of Veterans Administration (VA) and the Bureau of Indian Affairs (BIA) in the Department of Interior provided (and still provide) community mental health services directly in a number of locations.

Once the states became responsible for the distribution of the federal grant funds for mental health, many funds were shifted from the community mental health centers to community mental health services more responsive to the needs of the seriously mentally ill (Hudson 1983). The complex intergovernmental array of organizations involved made (and still makes) coordination difficult. In 1986, the State Comprehensive Mental Health Services Plan Act (PL 99-660) required the states to provide case management as a condition of receiving federal funding; it also mandated the development of state mental health plans. Various coordinating mechanisms were put in place to develop a wider base of service and support systems for the chronically mentally ill, including case management, multidisciplinary treatment teams, community support systems, integrated services, substate mental health authorities, and funding through pooling of resources of capitation (funding provided per individual served) (Dill and Rochefort 1989).

In California in 1989, AB 904 built upon the federal legislation. The result was the California Mental Health Master Plan, a blueprint for mental health services to the year 2000, which has continued to be updated. In 1991, the California legislature enacted the most sweeping reforms to mental health services since the LPS Act of 1969. The reforms, referred to as "realignment," included programmatic, governance, and fiscal changes. They became the groundwork for future reforms under California's Medicaid Managed Mental Health Plan. Under realignment, mental health was no longer funded as part of the state's general fund. The new funding approach sought to provide a stable funding source with growth based on economic factors. The amount of revenue dedicated to realignment was to equal the funds expended on the program in the prior year, but due to an economic recession, revenue actually fell 11 percent below projections. Not until 1995–1996 did the level of revenue attain the 1990–1991 level. Realignment was funded by a half-cent increase in sales tax, and by an increase in vehicle license fees. The funds were (and still are) collected by the state and maintained in a state trust. A statutory formula established how to distribute the money to counties for services, and statutes defined what the trust funds may be expended on.

The program reform established a common vision and philosophy of services, set target populations, established systems-of-care treatment approaches, put in place a minimum array of services, increased participation in decision making by consumers and advocates, set responsibility for program design at the local level, created incentives for restructuring at the state hospitals, and tied accountability to performance outcome measures. Three reports found realignment substantially successful. The California Mental Health Planning Council (CMHPC), a statutorily defined citizen's oversight commission; the California Mental Health Directors Association; and the California Legislative Analyst released these reports in 1991.

Advocacy systems to protect the rights of persons with mental illness were also put in place. Consumer groups for persons with mental illness grew (and are still growing), playing an increasingly important role in lobbying for funding and for changes in the mental health system. However, the approaches used and their availability varied community by community under a fragmented intergovernmental system (as is still the case).

Mental health care continued to be both inpatient and community based, but the site of inpatient care shifted from state institutions to general acute care hospitals in the community, with many people seen in psychiatric units in general acute care hospitals or in short-term public or private community inpatient facilities. Children began to be able to receive services in special community or residential treatment centers for children. The cost of inpatient care rose at all sites, but most sharply in general hospitals. Total costs were held in check because the length of stay decreased. Most of the decrease occurred in state mental hospitals and VA facilities, even though those facilities had (and still have) the longest stays (Kiesler and Sibulkin 1987). Community inpatient services were, as they still are, complemented by community outpatient services, such as the private practices of mental health professionals, family services agencies, community mental health centers, social clubs, day hospitals, halfway houses, group homes, assisted housing, and foster care. Agency models were diversifying, and there was a trend toward greater specialization, which has continued. Specialized services such as special education for emotionally disturbed children in the school system, forensic programs in the criminal justice system, and co-occurring (dual diagnosis) programs for people diagnosed with both mental illness and substance abuse problems. By the last decade of the twentieth century, community mental health services involved 2,965 organizations providing services to 3 million people, or 1.22 percent of the population (NIMH 1992, 37).

Funding issues remained critical. In an intergovernmental system each level of government tried to place more costs on another level of government. Many persons with mental disabilities continued to be transinstitutionalized into nursing homes and large group homes. States sought ways to receive federal funds for these facilities, particularly through Medicaid, which in turn looked for ways to limit costs by defining who is eligible for reimbursement. The U.S. Omnibus Reconciliation Act of 1987 required screening of new and current nursing home patients for mental illness to limit the use of nursing homes for persons with mental illness. The welfare program Supplemental Security

Income (SSI) remained a source of major funding for people whose mental disability left them permanently disabled.

In 1990 the Public Citizen Health Research Group and the National Alliance for the Mentally Ill released a report concluding that, as the century was drawing to a close, public psychiatric services were in near-total breakdown. Among their findings was that only one in five of the 2.8 million people with serious mental illness was receiving adequate care. The federal budget for mental health research in 1990 was $515 million, or only a fraction of the billions spent on AIDS, cancer, and heart disease (Torrey et al. 1990).

The report also found that not since the 1820s had so many people with mental illness been living untreated on the streets, in public shelters, and in jails. Many more were living in substandard boardinghouses or transient hotels. Mental health services were consuming approximately $20 billion per year, with an additional $3 billion in federal funds to create community mental health centers and more than $2 billion in federal funds and additional billions in state funds to train mental health professionals. Yet the report found that the system of services was not even minimally acceptable.

The report found eight current crises that demonstrated a near-total breakdown in public psychiatric services:

- There are more than twice as many (at least 150,000) people with schizophrenia and bipolar disorder living in public shelters and on the streets as there are in public mental hospitals.
- There are more people with bipolar disorder and schizophrenia in prison and jails (about 100,000) than in public mental hospitals.
- Increasing episodes of violence by seriously mentally ill persons are a consequence of not receiving treatment.
- Mental health professionals have increasingly left the public sector and gone to work in the private sector.
- Most community mental health centers have been failures.
- Funding of public services for individuals with serious mental illness is chaotic.
- An undetermined portion of public funds for services for people with mental illness is being stolen.

- Guidelines for serving people with mental illness are often created, at both the state and federal level, by administrators with no experience (Torrey et al. 1990).

Six proposals were made to improve services:

- Public mental health programs must serve people with serious mental illness as a priority, and if less than 75 percent of a program's resources are going to that group, its state and federal subsidies should be terminated.
- All mental health professionals should be required to donate, pro bono, one hour a week of work to public programs, and government-supported training programs for such professionals should include an automatic payback agreement.
- Because psychiatrists have largely left the public sector, psychologists, physician assistants, and nurse practitioners should be given special training and allowed to prescribe psychiatric medications.
- Funding of public services for persons with serious mental illness should be reorganized.
- Budgets of public mental illness programs should be examined for possible theft.
- All administrators of public programs for people with mental illness should spend at least one-half day each week working with people with mental illness.

Like its predecessors in 1986 and 1988, the survey used data from a variety of sources. The database was three times the size of the 1988 database, and sources included federal and state agencies, including for the first time directors of state departments of corrections reporting on mental illness services in state prisons, consumers and their family members, and experts in the field of mental illness services. Visits were made to thirty-five states. Data was compared among states and with what was considered to be the ideal services.

Vermont received the best score, followed by Rhode Island, New Hampshire, Connecticut, Ohio, and Colorado. Hawaii was in last place, with Idaho and Wyoming tied for forty-ninth. Montana and Mississippi tied for forty-seventh, and Texas and Nevada for forty-fifth. The state that showed the most improvement was Delaware. States moving backward were California, Florida, Indiana, and Pennsylvania (Torrey et al. 1990, 37–43).

The average state per capita spending on mental health was $42.11. Average state spending was 28 cents for every $100 of per capita income (Torrey et al. 1990, 43). Richer states tended to have better vocational rehabilitation, housing, and children's services but not necessarily inpatient or outpatient services. Northeastern states provided better services, while western states spent the least money and the least per capita income on mental health programs (Torrey et al. 1990, 48).

Special Diagnostic Issues

The 1999 Surgeon General's report on mental health care in the United States (U. S. Department of Health and Human Services, 1999) found that about 20 percent of the American population experience a diagnosable mental health condition each year, although as few as one-third receive treatment. Reasons given for so few being treated are stigma, financing, and access to providers.

A wide array of mental illnesses and disabilities appear in people throughout the community. These conditions include severe mental illnesses such as psychotic disorders like schizophrenia, affective disorders like bipolar disorder or severe clinical depression, and phobias like anxiety. Some people will experience one acute episode and will recover, never to suffer from that mental illness again. Other people will suffer from a chronic, severe mental disorder for the rest of their lives.

Schizophrenia affects about 2 million people in the United States in a year, or 1.6 percent of the population (Regier, Narrow, Rae, et al. 1993, 85). Affective disorders include major depression and bipolar disorder (manic depression). Major depression affects about 6.4 million people, or 3.5 percent of the population, with more women affected than men—6.6 percent of women and 3.3 percent of men in any year. About 1.1 million people, or 1 percent of the population, are affected by bipolar disorder (Robins, Locke, and Regier 1991, 34). The number of people receiving treatment for depression has increased dramatically since 1987, according to a study at Columbia University. The study found that between 1987 and 1997, the percentage of Americans being treated for depression tripled nationwide from 0.7 percent to 2.3 percent. The study also found that the number of visits to physicians for depression decreased from about twelve per patient to eight. Large numbers of patients receive treatment from primary care physicians and others outside the mental health field (California Healthline 2002). Anxiety disorders, including obsessive-compulsive disorder, panic disorder, phobia, and post-traumatic stress disorder, affect many Americans each year. Obsessive-compulsive disorder affects about 2.9 million people (1.6 percent of the population) and phobia 16.2 million people (8.8 percent of the population). (National Advisory Mental Health Council 1993, 1446).

The definition of serious mental illness (SMI) has evolved over time, and there is increased recognition that people with SMI are a heterogeneous group with different diagnoses, levels of disability, and durations of disability. In a study of serious mental illness and disability in adults, the National Center for Health Statistics defined SMI as "any psychiatric disorder present during the past year that seriously interfered with one or more aspects of a person's daily life" (Barker et al. 1992). Conflict sometimes arises between stakeholders as to whether the severely mentally ill should be given the highest priority in treatment.

Mental disorders generate an immense public health and economic burden. The World Health Organization, in collaboration with the World Bank and Harvard University, has determined the "burden of disability" associated with the wide range of mental disorders suffered by people throughout the world. In their study, *Global Burden of Disease,* they report that the impact of mental illness on overall health and productivity in the United States and around the world is profoundly underrecognized. Lost productivity and disability insurance payments due to illness or premature death account for $74.9 billion a year (U.S. Department of Health and Human Services 1999). In established market economies like the United States, mental illness is on a par with heart disease and cancer as a cause of disability (Murray and Lopez 1996).

References

Bachrach, L. L. (1979). Planning Mental Health Services for Chronic Patients. *Hospital and Community Psychiatry* 30:387–393.

Bandura, A. (1978). On Paradigms and Recycled Ideologies. *Cognitive Therapy and Research* 2:79–103.

Bardach, E. (1977). *The Implementation Game.* Cambridge, MA: MIT Press.

Barker, P. R., R. W. Manderschied, G. E. Hendershot, S. S. Jack, C. Schoenborn, and I. Goldstrom (1992). Serious Mental Illness and Disability in the Adult Household Population: United States, 1989. *Advance Data* 218 (September 16): 1–11.

Barter, J. T. (1983). California Transformation of Mental Health Care: 1957–1982. In *Unified Health Systems: New Directions for Mental Health Services,* ed. J. A. Talbott. San Francisco: Jossey-Bass.

California Council of Community Mental Health Agencies (CCCMHA) (n.d.). The Development of California's Publicly Funded Mental Health System. Prepared by S. Naylor Goodwin and R. Selix. http://www.cccmha.org/ (accessed September 20, 2006).

California Department of Institutions (1939). Statistical Report 1938–1939. Sacramento: California Department of Institutions.

California Healthline. Depression Treatment Up, with More Using Drug Therapy. California Health Care Foundation (January 9, 2002). http://www.californiahealthline.org/index.cfm?Action=itemPrint&ItemID=90897 (accessed August 11, 2006).

Caplan, R. B. (1969). *Psychiatry and the Community in Nineteenth-Century America: The Recurring Concern with Environment in the Prevention and Treatment of Mental Disorder.* New York: Basic Books.

Castro, J. (1993). What Price Mental Health? *Time* 14 (22) (May 31):59–60.

Dain, N. (1964). *Concepts of Insanity in the United States, 1789–1865.* New Brunswick, NJ: Rutgers University Press.

Deutsch, A. (1949). *The Mentally Ill in America.* 2nd ed. New York: Columbia University Press.

Dill, A., and D. Rochefort (1989). Coordination, Continuity and Centralized Control: A Policy Perspective on Service Strategies for the Chronic Mentally Ill. *Journal of Social Issues* 45:145–159.

Final report to the California Legislature of the Senate Select Committee on Proposed Phaseout of State Hospital Services (1974). Sacramento: California Senate, March 15.

Foley, H. A. (1975). *Community Mental Health Legislation: The Formative Process.* Lexington, MA: Heath.

Fox, W. (1978). *So Far Disordered in Mind: Insanity in California 1870–1930.* Berkeley: University of California Press.

Friedman, L. (1985). Caring for the Mental Patient in the General Hospital: A Comparison of Hospitals with and without Psychiatric Units. Ph.D. diss. University of Pittsburgh.

Goldstrom, I. D., and R. W. Manderschied (1982). The Chronically Mentally Ill: A Descriptive Analysis from the Uniform Client Data Instrument. *Community Support Service Journal* 2:4–9.

Grob, G. N. (1966). *The State and the Mentally Ill: A History of Worcester State Hospital in Massachusetts, 1830–1920.* Chapel Hill: University of North Carolina Press.

Grob, G. N. (1973). *Mental Institutions in America: Social Policy to 1875.* New York: Free Press.

Grob, G. N. (1983). *Mental Illness and American Society, 1875–1940.* Princeton, NJ: Princeton University Press.

Grob, G. N. (1987). The Forging of American Mental Health Policy in America: World War II to New Frontier. *Journal of the History of Medicine* 42:410–446.

Group for the Advancement of Psychiatry (1978). *The Chronic Mental Patient in the Community.* New York: Mental Health Materials Center.

Hudson, C. G (1983). An Empirical Model of State Mental Health Spending. *Social Work Research and Abstracts* 23:312–322.

Hudson, C. G (1993). The United States. In *International Handbook on Mental Health Policy,* ed. D. Kemp, 413–446. Westport, CT: Greenwood Press.

Kennedy, J. F. (1964). Special Message to the Congress on Mental Illness and Mental Retardation. *Public Papers of the Presidents of the United States: John F. Kennedy, 1963.* Washington, DC: U.S. Government Printing Office.

Kiesler, C. A. (1980). Mental Health Policy as a Field of Inquiry for Psychology. *American Psychologist* 35:1066–1088.

Kiesler, C. A. (1982). Mental Hospitals and Alternative Care: Noninstitutionalization as Potential Public Policy for Mental Patients. *American Psychologist* 37:349–360.

Kiesler, C. A. (1983). Psychology and Mental Health Policy. In *The Clinical Psychology Handbook,* ed. M. Hersen, A. D. Kazdin, and A. S. Bellack. New York: Pergamon Press.

Kiesler, C. A. (1987). The Guilds: Their Organization, Members and Potential Contributions. In *The Future of the Mental Health Services: Coping with Crisis,* ed. L. J. Duhl and N. A. Cummings. New York: Springer.

Kiesler, C. A., and A. E. Sibulkin (1987). *Mental Hospitalization: Myths and Facts about a National Crisis.* Newbury Park, CA: Sage Publications.

Lamb, H. R., and V. Goertzel (1977). The LongTerm Patient in the Era of Community Treatment. *Archives of General Psychiatry* 34:679–682.

Lewis, D.A., W. R. Shadish Jr., and A. J. Lurigio (1989). Policies of Inclusion and the Mentally Ill: Long-Term Care in a New Environment. *Journal of Social Issues* 45:173–186.

McGovern, C. M. (1985). *Masters of Madness: Social Origins of the American Psychiatric Profession*. Hanover, NH: University Press of New England.

Moltzon, S., H. Gurevitz, M. Rappaport, et al. (1986) The Psychiatric Health Facility. *Hospital and Community Psychiatry* 37:1131–1135.

Mosher, L. R. (1983). Alternatives to Psychiatric Hospitalization: Why Has Research Failed to be Translated into Practice? *New England Journal of Medicine* 309:1579–1580.

Murray, C. J. L., and A. S. Lopez (1996). *The Global Burden of Disease*. Cambridge, MA: Harvard University Press.

National Advisory Mental Health Council (1993). Health Care Reform for Americans with Severe Mental Illnesses: Report of the National Advisory Mental Health Council. *American Journal of Psychiatry* 150:1437–1446.

National Association for the Mentally Ill (NAMI) (2006). *Grading the States 2006: A Report on America's Health Care System for Serious Mental Illness*. http://www.nami.org/gtstemplate.cfm?section=gradingthestates (accessed November 25, 2006).

National Institute of Mental Health (NIMH) (1992). Mental Health, United States, 1992. DHHS publication no. ADM 90–1708. Washington, DC: Superintendent of Documents, Government Printing Office.

Nutting, J. (1902). The Poor, the Defective, and the Criminal. In *State of Rhode Island and Providence Plantations at the End of the Century*, ed. E. Field. Boston: Mason.

Regier, D.A., W. Narrow, D. S. Rae, et al. (1993). The De Facto U.S. Mental and Addictive Disorders Service System: Epidemiologic Catchment Area Prospective 1-year Prevalence Rates of Disorders and Services. *Archives of General Psychiatry* 50:85–94.

Report of the Mental Health Legislative Work Group: A Model for California Community Mental Health Programs: Phase II (1981). Sacramento: Mental Health Legislative Work Group.

Roberts, N. (1967). *Mental Health and Mental Illness*. London: Routledge and Kegan Paul.

Robins, L. N., B. Z. Locke, and D. A. Regier (1991). An Overview of Psychiatric Disorders in America. In *Psychiatric Disorders in America: The Epidemiologic Catchment Area Study*, ed. L. N. Robins and D. A. Regier. New York: Free Press.

Rochefort, D. A. (1993). *From Poorhouses to Homelessness: Policy Analysis and Mental Health Care*. Westport, CT: Auburn House.

Rochefort, D. A. (1997). *From Poorhouses to Homelessness: Policy Analysis and Mental Health Care.* 2nd ed. Westport, CT: Auburn House.

Rodgers, J. E. (1992). *Psychosurgery: Damaging the Brain to Save the Mind.* New York: HarperCollins.

Rothman, D. J. (1971). *The Discovery of the Asylum: Social Order and Disorder in the New Republic.* Boston: Little, Brown.

Rothman, D. J. (1980). *Conscience and Convenience: The Asylum and its Alternatives in Progressive America.* Boston: Little, Brown.

Rubins, J. L. (1971). The Community Mental Health Movement in the United States, Circa 1979. Pt. 1. *American Journal of Psychoanalysis* 31:68–79.

Shutts, D. (1982). *Lobotomy, Resort to the Knife.* New York: Van Nostrand Reinhold.

Torrey, E. Fuller, Karen Erdman, Sidney M. Wolfe, and Laurie M. Flynn (1990). *Care of the Seriously Mentally Ill: A Rating of State Programs.* Washington, DC: Public Citizen Health Research Group and National Alliance for the Mentally Ill.

U.S. Congress, Senate, Subcommittee on Health of the Committee on Labor and Public Welfare (1963). *Hearings on S. 755 and S. 756, 88th Cong., 1st Sess.* Washington, DC: U.S. Government Printing Office.

U.S. Department of Health and Human Services (1999). *Mental Health: A Report of the Surgeon General—Executive Summary.* Rockville, MD: HHS, SAMHSA, CMHS, NIH, NIMH.

Warren, C. A. B. (1982). *The Court of Last Resort.* Chicago: University of Chicago Press.

Zilboorg, A., and G. W. Henry (1941). *A History of Medical Psychology.* New York: Norton.

Zusman, J. (1975). The Philosophic Basis for a Community and Social Psychiatry. In *An Assessment of the Community Mental Health Movement,* ed. W. F. Barton and C. J. Sanborn. Lexington, MA: D.C. Heath.

2

The Twenty-First Century: Problems, Controversies, and Solutions

As was the case in the twentieth century, the United States has no national mental health system. Each state has its own distinctive system. This approach allows for adjusting programs to the unique characteristics of different states and communities, but has the disadvantage of creating disparities and differences in levels of deinstitutionalization and levels of community services. The private, nonprofit, and public sectors all play major roles in the delivery of services to people with mental illness. The system remains two-tiered, with lower-income people relying on the public sector and higher-income people on the private sector. People with insurance or sufficient income can access private mental health providers, from private psychotherapists, to general hospital psychiatric units in private and nonprofit hospitals, to private psychiatric facilities. The public mental health system remains the provider of last resort for people needing mental health services. However, there is a trend for more of these services being contracted out to the private sector rather than being provided by the public sector. The missions of most state mental health agencies focus resources on people with the most severe and persistent mental illnesses, such as bipolar disorder and schizophrenia.

The states remain the critical players in the development and maintenance of the public mental health system. "In fact, mental health, more than any other public health or medical discipline, is singled out for exclusion and discrimination in many federal programs because it is considered to be the principal domain of the states" (Urff 2004, 84). In most states, the mental health system is administered by a state mental health agency. This agency may be an independent department, but is most often an agency within a larger department, usually health or social services. About one-third of the states combine mental health and services for people with developmental disabilities in the same agency. About half of the states combine mental health and substance abuse services in the same agency (Lutterman and Shaw 2000). Almost all states still have state-operated psychiatric hospitals. All state mental health agencies provide funding to community service providers, and many state mental health agencies license private and nonprofit community service providers and monitor their performance. Funding is provided by reimbursement for specific services, contracts for specific programs, and formula grants. As mentioned above, the dividing line between the public and private delivery system is becoming more blurred (Urff 2004). As states downsize and close public psychiatric hospitals, services provided by private and nonprofit organizations take on increasing importance.

Some state mental health agencies administer mental health benefits under the state's Medicaid plan, especially if they have adopted Medicaid managed care; others pass this function on to counties. Many people who would have been clients of the public mental health system are now seen in the private sector when private providers are able to bill Medicaid or Medicare for reimbursement.

In the twenty-first century, the history of modern mental health care continued to be complex and cyclical. There was a tug and pull between many of the viewpoints about mental illness. Should people with mental illness be free to manage their lives as they saw fit or should there be social control by the government? Was mental illness physical, environmental, or both? Was mental illness just a part of physical illness? In this chapter we will look at some of the major issues and problems still important in the twenty-first century.

Defining Mental Illness

Definitions of mental disabilities have changed over time. The basis of definition has moved from culturally deviant; to possession by the devil; to sickness; to disturbances in developmental, familial, genetic, and biochemical processes. The World Health Organization (WHO) (1946) defines health as "A state of complete physical, mental, and social well-being." The U.S. Department of Health and Human Services in *Healthy People 2010* (2000) defines mental disorders as "health conditions that are characterized by alterations in thinking, mood, or behavior (or some combination thereof), which are associated with distress and/or impaired functioning and spawn a host of human problems that may include disability, pain or death. *Mental illness* is the term that refers collectively to all diagnosable mental disorders"(p. 36). It defines mental health as "a state of successful performance of mental function, resulting in productive activities, fulfilling relationships with other people, and the ability to adapt to change and to cope with adversity. Mental health is indispensable to personal well-being, family and interpersonal relationships, and contribution to community or society"(p. 36). In Canada, mental health is defined as the capacity of the individual, the group, and the environment to interact with one another in ways that promote subjective well-being, the optimal development and use of mental abilities (cognitive, affective, and relational), the achievement of individual and collective goals consistent with justice, and the attainment and preservation of conditions of fundamental equality (Health and Welfare Canada, 1988).

There are two systems of diagnosis widely used in the United States, one found in *The Diagnostic and Statistical Manual of Mental Disorders* (*DSM-IV*) and the other in the *Manual of the International Statistical Classification of Disease, Injuries and Causes of Death*, known as the *ICD-10*. The *DSM* first appeared in 1952. It provides the most commonly used resource for U.S. mental health practitioners making diagnostic decisions. It was developed by the American Psychiatric Association to provide a detailed description of all categories of mental illness and is based on a combination of research field trials and committee votes.

The growing consumer movement is sometimes conflicted on how mental illness should be defined. Different consumer groups place emphasis on different populations with mental health problems. NAMI is campaigning to establish mental illnesses as brain disorders. It believes that the definition change will lead to more resources (Jacobson 2004). Other interest groups identify environmental factors, including violence and poverty, as causes of mental disorders and as preventing recovery. These groups emphasize a broader range of priority populations and issues. Specialized attention is being focused on children, the elderly, co-occurring disorders with substance abuse, and AIDS, which in a sizable number of cases includes dementia.

Co-occurring Disorders (Dual Disorders)

The co-occurrence of addictive disorders among persons with mental disorders is gaining increasing attention. Co-occurrence has also been called dual disorder or dual diagnosis. Among adults aged eighteen years and older with a lifetime history of any mental disorder, 29 percent have a history of an addictive disorder. Of those with an alcohol disorder, 37 percent have had a mental disorder. Among those with other drug disorders, 53 percent have had a mental disorder (U.S. Department of Health and Human Services 1999). About 3 percent of the population aged eighteen years and older has been diagnosed with a co-occurring mental and addictive disorder in one year. Of those with a serious mental illness, 15 percent have both types of disorder in one year, and of those with a severe and chronic disorder, 27 percent have both mental and addictive disorders (Kessler, Berglund, Zhao, et al. 1996). Some addiction in people with mental disorders can be ascribed to people self-medicating themselves in trying to deal with their mental disorder. For example, people with depression may try to make themselves feel better by using alcohol or other drugs.

The standard treatment for co-occurring disordered individuals is "parallel treatment," in which the person is treated by separate agencies specializing in treating either mental illness or substance abuse. The care is rarely coordinated. In practice, individuals may be excluded from one or both systems. Research indicates that outcomes in parallel treatment are poor (Drake, Essock, Shaner, et al. 2001).

Integrated treatment (IT) has been advocated for treating dual disorder individuals (Drake, Essock, Shaner, et al. 2001). In IT a single clinician or team provides both mental health and substance abuse treatment in a coordinated manner. IT and active community treatment (ACT) have produced better outcomes in terms of hospitalization, housing, mental health, and sometimes substance abuse, but not in relationship to criminal behavior in dual-disordered clients (Calsyn, Yonker, Lemming, et al. 2005).

Suicide

Suicide, a major public health problem in the United States and many countries, occurs most frequently as a consequence of a mental disorder. Suicide may be the worst outcome of a mental disorder. Suicide rates have become very high among the young. Legislation was introduced in Congress in 2004 to provide funding to the states for more suicide prevention programs.

Risk for engaging in suicidal behaviors differs by gender. A history of physical or sexual abuse appears to be a risk factor for suicide attempts in both men and women (National Center for Health Statistics 1991). Women attempt suicide more often than men, but men are more likely to complete the suicide attempt (Moscicki 2001). Completed suicide is on average four and a half times higher for men than women (Centers for Disease Control 1999). This suicide gender gap begins in adolescence and grows through middle and later years. In 1999, *The Surgeon General's Call to Action to Prevent Suicide* provided a plan to increase awareness and prevent suicide in the United States.

Who Are the Mentally Ill?

Mental disorders occur across the lifespan, affecting all people regardless of race, ethnicity, gender, or educational and socioeconomic status. In the mid-1990s in the United States, approximately 40 million people aged eighteen to sixty-four years (22 percent of the population) had a diagnosis of mental disorder alone (19 percent of the population) or of a co-occurring mental and addictive disorder (3 percent of the population) in the past year (Kessler, McGonagle, Zhao, et al. 1994; Regier, Narrow, Rae, et al. 1993). The World Health Organization has estimated that approximately 450 million people worldwide have mental and behavioral disorders, and mental disorders account for 25 percent of all disability in major industrialized countries (World Health Organization 2001, 7). The most severe forms of mental disorders have been estimated to affect between 2.6 percent and 2.8 percent of adults aged eighteen years and older during any one year (Kessler, Berglund, Zhao, et al. 1996; National Advisory Mental Health Council 1993). About 15 percent of adults receive help from mental health specialists, while others receive help from general physicians. The majority of people with mental disorders do not receive treatment, and 40 percent of people with a severe mental illness do not look for treatment (Regier, Narrow, Rae, et al. 1993). Dr. Ronald C. Kessler, principal investigator of the National Comorbidity Survey Replication study, and colleagues determined that about half of Americans will meet the criteria for a *DSM-IV* diagnosis of a mental disorder over the course of their lifetime, with first onset usually in childhood or adolescence. Based on their analysis, lifetime prevalence for the different classes of disorders were anxiety disorders, 28.8 percent; mood disorders, 20.8 percent; impulse control disorders, 24.8 percent; substance use disorders, 14.6 percent; and any disorder, 46.4 percent.

Median age at onset is much earlier for anxiety and impulse control disorders (11 years for both) than for substance use (20 years) and mood disorders (30 years), and half of all lifetime cases start by age 14, and three-fourths by age 24 (Romano 2005).

One in ten children and adolescents suffers from mental illness severe enough to result in significant functional impairment (National Advisory Mental Health Council 2001). At least one in five children or adolescents between the ages of nine and seventeen years has a diagnosable mental disorder in a given year (Shaffer, Fisher, Dulcan, et al. 1996). Behavioral and mental disorders and serious emotional disturbances (SEDs) in children and adolescents can lead to school failure, violence, suicide, or alcohol or illicit drug use, and about 5 percent of children and adolescents are extremely impaired by mental, emotional, and behavioral disorders (Brandenberg, Friedman, and Silver 1990; Foley, Carlton, and Howell 1996; Friedman, Katz-Leavy, Manderschied, et al. 1996; Taylor, Chadwick, Heptinstall, et al. 1996). The landmark document *Mental Health: A Report of the Surgeon General* released in 1999 by the U.S. Public Health Service included a chapter focused on the mental health needs of children. The report marked a turning point in the public's interest in mental health, clearly documenting the public health need for an effective mental health service and highlighting the scientific advances made. A continuation of that work was *A Report of the Surgeon General's Conference on Children's Mental Health: A National Action Agenda* in 2000, which provided a blueprint for children's mental health research, practice, and policy. In 2001, *Blueprint for Change: Research on Child and Adolescent Mental Health* was released by the National Advisory Mental Health Council's Workgroup on Child and Adolescent Mental Health Intervention, Development, and Deployment. The report established three priority areas: research in basic science and the development of new interventions, intervention development that would allow moving from efficacy to effectiveness, and intervention deployment that would allow moving from effectiveness to dissemination. To mark this new generation of research, the workgroup made recommendations in three broad areas: interdisciplinary research development in child and adolescent mental health with child and adolescent interdisciplinary research networks, interdisciplinary research training in child and adolescent mental health, and recommendations for program development in specific research areas including neuroscience, behavioral science, prevention, psychosocial interventions, psychopharmacology, combined interventions and services, dissemination of research, and improvement of the child and adolescent mental health system.

It is estimated that about 25 percent of people aged sixty-five years and older experience some mental disorder, including depression, dementia, anxiety, and substance abuse, that is not part of normal aging (Ritchie and Kildea 1995). Alzheimer's disease affects 8 to 15 percent of people over sixty-five and probably accounts for 60 to 70 percent of all cases of dementia. It is one of the leading reasons for nursing home placements (Ritchie and Kildea 1995; Lebowitz, Pearson, and Cohen 1998). With the aging of the American population, the urgency of addressing the mental health needs of the elderly is growing.

Treatment

The disease model of mental illness has remained important. For the seriously mentally ill, it places a focus on finding and treating the causes of the emotional, behavioral, and/or organic dysfunction, with an approach based on diagnosis, treatment, and cure. The treatment focus is on short-term inpatient care, with the emphasis on medications and the ability to function in the community. Various services are provided to assist maintenance in the community, including housing, employment, and social services.

The mental health approach to people with less serious emotional disorders is focused on outpatient treatment, often through prescription of medication by a general practitioner. Primary care physicians provide at least 40 percent of mental health care. Some people are treated through community psychologists, family counselors, or clinical social workers with training in psychotherapy. Employee assistance programs help employees in the workplace with assessment of mental health issues and referral to appropriate treatment sources. Those who are less seriously mentally ill are sometimes referred to in a derogatory way as the "worried well." When resources are short, conflict can arise, as those who speak on behalf of the seriously and chronically mentally ill do not wish to see resources expended on the less seriously ill. However, many people at one time need assistance with a mental health problem in their lives, and failure to address their problems can lead to significant costs to society and even to suicide.

In California, licensed mental and behavioral health care workers total approximately 63,000, with 37 percent being marriage and family therapists, 22 percent licensed clinical social workers, 18 percent psychologists, 15 percent psychiatric technicians, 8 percent psychiatrists, and 1 percent advance nurses with advanced training in psychiatry or mental health (McRee, Dower, Briggance, et al. 2003). National estimates indicate that over 70 percent of social workers and psychologists are non-Hispanic whites and 65 percent are women. Mental and behavioral health care workers provide care in a variety of settings, including state psychiatric hospitals, private or nonprofit psychiatric hospitals, psychiatric services in acute care hospitals and specialty hospitals, county mental health programs, community clinics, private practice settings, criminal justice and correctional facilities, and schools (McRee et al. 2003). There is a shortage of psychiatrists in public mental health facilities.

The shortage of psychiatrists for children is especially critical. The Center for Mental Health Services estimates that at least 5 percent of U.S. children and adolescents have acute mental disorders (Shortage of Psychiatrists 2006, 8B). With declining stigmas regarding mental health problems, more parents are willing to seek help. The shortage of child psychiatrists is connected to two factors: the extra two years of training required on top of four years of medical school and three years of general psychiatry, and the reimbursement rate, which does not reflect the extra time that psychiatrists must take to interview parents, teachers, and others. The American Academy of Child and Adolescent Psychiatrists (AACAP) estimated the number of child psychiatrists at 7,000 (Shortage of Psychiatrists 2006, 8B), while the U.S. Bureau of Health Professions projected that there would be about 8,300 by 2020, only two-thirds of the estimated 12,600 needed (Shortage of Psychiatrists 2006, 8B). AACAP found in 2003 that there was on average only one child psychiatrist for every 15,000 children and adolescents. West Virginia had only 1.2 psychiatrists per 100,000 children and adolescents (Shortage of Psychiatrists 2006, 8B). Some states are using *tele-psychiatry* (using video over the Internet), which allows physicians in rural areas to consult with urban or university child psychiatrists. The lack of child psychiatrists results in children being seen by family doctors, pediatricians, and child psychologists, who have far less training. The National Alliance on Mental Illness has also reported that parents who cannot find or afford private psychiatric care are told to give up custody of their child to the state in order to get treatment (National Alliance for the Mentally Ill 1999).

Studies suggest that the prevalence of mental illness is about equal in urban and rural areas, but access to services is much more difficult in rural areas. Ninety-five percent of the nation's rural counties do not have access to a psychiatrist, 68 percent do not have access to a psychologist, and 78 percent lack access to social workers. According to the federal Center for Mental Health Services, about 8 million people lack ready access to a mental health professional (California Healthline 2000).

Research and Prevention

Targeted preventive interventions, implemented according to scientific recommendations, have great potential to reduce the risk for mental disorders. Social and behavioral research is beginning to explore the concept of resilience to identify strengths that may promote health and healing. Resilience involves the interaction of biological, psychological, and environmental processes. Identifying and promoting resilience may make it possible to design effective programs that draw on this internal capacity.

There is also an increased awareness in public health of the impact of stress, its prevention and treatment, and the need for improving coping skills. Stress may be experienced by anyone and is a clear demonstration of mind-body interaction. Coping skills, acquired throughout the lifespan, are positive adaptations that affect the ability to manage stressful events. Research can identify the public health burden of stress and identify ways to prevent it through environmental or individual strategies.

Research-based treatments provide an unprecedented opportunity to achieve a major reduction in the burden of disease associated with mental illness, and mental health services can be enhanced through research findings. Scientists have developed, tested, and structured preventive interventions for depression and other disorders in high-risk children. Preventive interventions can decrease risk of onset or delay onset of some disorders (U.S. Department of Health and Human Services 2000). It has been noted that children and adolescents with mental illnesses often do not become substance abusers until after the mental illness becomes apparent (Christie, Burke, Regier, et al. 1988). This may create a window of opportunity to treat children and adolescents before they have a co-occurrence.

With the growth in life expectancy and the aging of the baby boom, the mental health system will be faced with a growing population who may experience mental disorders of late life. This trend will present society with unprecedented challenges to develop effective preventive services for mental health in this population. Recognition is growing that depression and certain cognitive losses are not the inevitable consequences of aging.

One of the most controversial and divisive issues in mental health policy is whether people diagnosed with mental illness should be treated against their will. Outpatient commitment involves requiring people with psychiatric diagnoses who are living in the community to accept mandatory treatment including mandatory use of prescription drugs. A failure to comply can lead to inpatient commitment.

Consumer Choice and Involuntary Treatment

As far back as the 1930s, there have been documented instances of people with mental illness coming together for mutual support. What is now known as the consumer, or survivor, movement, however, is generally recognized as having begun in the early 1970s, when several small groups of people who had been involved in the mental health system began to meet together in several cities. They talked about their experiences and began to develop agendas for change. They drew from the antipsychiatry movement and the writings of Thomas Szasz, Erving Goffman, Michel Foucault, and others who viewed mental illness as not a disease of the mind but as a method of social control used against those who were deviant, their marginalization being based on gender, sexual orientation, poverty, or antiestablishment actions. This movement was patterned after the civil rights and women's movements. Its agenda for social change emphasized self-help and advocacy.

In 1867, the first laws regulating involuntary commitment were enacted in Illinois, following lobbying by Elizabeth Packard, whose minister husband had her committed after she disrupted his service by declaring she was the Holy Spirit. She was set free. Involuntary treatment is one of the factors that distinguish treatment of persons with mental illnesses from persons with physical illnesses. Involuntary treatment is extremely rare for physical illnesses and occurs most commonly in cases where the patient is unconscious or unable to communicate. Involuntary treatment is common in inpatient mental health treatment, and the number of states providing some type of involuntary community treatment has been increasing rapidly, moving from thirty-four states to forty-one states. Involuntary outpatient commitment (OPC) is a civil justice procedure intended to enhance compliance with community mental health treatment, to improve functioning, and to reduce recurrent dangerousness and hospital recidivism.

Involuntary treatment is an issue that engages the attention of not only mental health consumers, but family groups, providers, citizen advocacy groups, and law enforcement. On one side of the issue are those who would outlaw the use of force and coercion completely to protect individuals from dangerous interventions and abuse. They believe forced treatment violates basic civil and constitutional rights and erodes self-determination. They also believe forced treatment can lead to distrust and an avoidance of voluntary treatment. Those favoring involuntary treatment are found along a continuum, ranging from those who believe such treatment is justified only under extreme situations, when people are demonstrably dangerous to themselves or others, to those who believe involuntary treatment is acceptable based on a broad set of criteria.

The mental health consumer movement generally sees force and coercion as indicative of failed treatment and counterproductive. The movement is largely united in opposition to involuntary outpatient commitment. The National Summit of Mental Health Consumers and Survivors established a Force and Coercion Plank that reached consensus on the following:

- Outpatient commitment would not be necessary if there were appropriate community services available.
- Forced treatment drives people away from seeking voluntary treatment.
- Studies have shown that outpatient commitment has no positive value; what does make a difference is getting appropriate services.
- It is cheaper to put money into community services than to put it into the enforcement of outpatient commitment laws.
- People diagnosed with mental illnesses should have a voice in their own treatment.
- Choice is essential for recovery.
- Using violence as an argument for expanding forced treatment and outpatient commitment is wrong; every study shows that, absent drugs and alcohol, people with mental illness are no more violent than any other group of people. (National Mental Health Consumers' Self-Help Clearinghouse 1999)

Coalitions including mental health consumer groups have recently lobbied against involuntary outpatient commitment bills in Maryland, Iowa, Massachusetts, and California. The Bazelon Center, the American Civil Liberties Union, and others have argued strenuously for civil rights and treatment by informed consent only.

But there are those, such as E. Fuller Torrey of the Treatment Advocacy Center, who believe that many people with schizophrenia or bipolar disorder have brain damage that impairs their ability to recognize their own illnesses. They see measures such as involuntary community-ordered treatment as necessary and argue that protecting the civil liberties of a seriously mentally ill person living on the streets in fact devalues that person (Treatment Advocacy Center 2002).

Perhaps the most important argument used to justify involuntary outpatient commitment is that it is necessary to prevent violence on the part of the mentally ill, since mental disorder may be a significant risk factor in the occurrence of violence, and many episodes of violence are preventable.

California passed legislation nearly thirty years ago that set a standard of dangerousness to self or others before someone with a mental illness could be required to accept treatment. California and Connecticut also required state agencies to inform the public about advanced directives for people with mental illness and to make them more available at critical times, such as when a person was discharged from a psychiatric hospital. Advance directives allowed people with mental health problems to designate in an advance health directive who they wanted to make health decisions for them when they were not able to make those decisions themselves.

To prevent people with mental disabilities from falling into the criminal justice system, communities nationwide are seeking ways of treating people in the community. On a small scale this is being done by the development of mental health courts that place a priority on involuntary treatment while using punishment as a backup to enforce compliance, and these will be discussed later. But for those with mental disabilities who have not fallen into the criminal justice system, the question remains as to whether involuntary treatment should be provided in the community.

The incident that sparked legislation in California was the killing of nineteen-year-old high school valedictorian Laura Wilcox by a forty-one-year-old northern California man suffering from paranoia. His family had tried to persuade him to accept treatment and medication, but he refused. On January 10, 2001, Scott Harlan Thorpe shot to death Laura Wilcox, who was at that time a worker inside the county mental health office in Nevada County.

Assemblywoman Helen Thomson, a Democrat from Davis and a former psychiatric nurse, wrote AB 1421 during her last legislative session before term limits ended her Assembly career. The legislation was modeled after a New York law known as Kendra's Law, which was passed in 1999, eight months after a Manhattan office worker, Kendra Webdale, thirty-two, was pushed to her death in front of a subway car by Andrew Goldstein, a thirty-one-year-old schizophrenic man off his medication. Kendra's Law was based on a carefully controlled pilot program at Bellevue Hospital in New York City that tested the feasibility of outpatient civil commitment. When incentives and disincentives were used in managing more than 150 cases of patients with violent and disordered behaviors, there was only one minor violent incident from 1995 to 2000 (Stavis 2003).

California's AB 1421 legislation was clinically focused and attempted to provide treatment to a wider range of people in the community, who if they did not receive treatment would eventually require hospitalization. The legislation permitted a judge to order outpatient treatment for people who did not realize the gravity of their illness. AB 1421 was written to guarantee due process. The legislation also required consultation with family and mental health professionals.

Opponents showed up at legislative hearings wearing upside-down triangles, the symbol worn by mental patients in Nazi concentration camps. They denounced Assemblywoman Thomson, calling her Nurse Ratchet, the character who embodies the repressiveness of the mental health system in Ken Kesey's *One Flew over The Cuckoo's Nest.*

On June 19, 2002, the Senate Health and Human Services Committee passed AB 1421 unanimously. Seven amendments made to the bill were not consented to by Assemblywoman Thomson. Six of the amendments further limited the legislation, including requiring public defenders be present at any commitment hearing and requiring county mental health directors to review each petition to ensure that severely mentally ill people would not be sent to court frivolously. The seventh amendment negated much of the original intent of the legislation by restricting eligibility for the new program to people who met California's current 5150 standard, which requires that those covered by it be in imminent danger of harming themselves or others. A system for 5150 hospitalizations already existed. The Senate Judiciary Committee then passed the bill with amendments that returned the bill to something closer to its original form.

The National Association for the Mentally Ill (NAMI) was a major supporter of the bill. Most NAMI members believed the legislation would give courts and doctors the authority to treat those unable to request treatment or recognize their illness.

The California Network of Mental Health Clients organized the opposition. Opponents included groups representing disabled rights, homeless rights, and the antipsychiatry movement. They believed a person's right to refuse treatment should not be taken away. Client organizations also believed that clients' right to make their own decisions was crucial to recovery. They supported instead better options for services, including funding for housing and peer-to-peer and other support organizations.

AB 1421 has been enacted as the Assisted Outpatient Treatment Demonstration Project Act of 2002 to create an assisted outpatient treatment program for any person who is suffering from a mental disorder and meets certain criteria. The program will operate in counties that choose to provide the services. Each county operating an outpatient treatment program must provide data to the Department of Mental Health, which must make a report to the legislature. The demonstration project will end on January 1, 2008.

A petition meeting specific requirements must be filed alleging the necessity of treatment. The patient has certain rights, which must be respected, and specified hearing procedures must be followed. Settlement agreements may be made in lieu of the hearing process. If the person who is the subject of the petition fails to comply with outpatient treatment, despite efforts to solicit compliance, a licensed mental health treatment provider may request that the person be placed under a seventy-two-hour hold for an involuntary commitment.

The program involves the delivery of community-based care by multidisciplinary teams of mental health professionals with staff-to-client ratios of not more than one to ten. Additional services are to be provided for persons with the most persistent and severe mental illnesses.

The legislation is expected to have a very significant direct impact for people with mental health disabilities and their families in counties that choose to implement a demonstration project. The impact on vendors, providers, and direct care staff in the participating counties is not yet clear, but the results of a study in New York City may be an indication of what will happen.

An effectiveness evaluation of three years of the outpatient commitment pilot program established in 1994 at Bellevue Hospital in New York City studied a total of 142 participants who were randomly assigned, 78 receiving court-ordered treatment, which included enhanced services, and 64 receiving the enhanced-service package only. Between 57 percent and 68 percent of the subjects completed interviews at one, five, and eleven months after hospital discharge. Outcome measures included rehospitalization, arrest, quality of life, symptomatology, treatment noncompliance, and perceived level of coercion. No significant differences were found between the two groups on all the major outcome measures, and no subject was arrested for a violent crime. Eighteen percent of the court-ordered group and 16 percent of the control group were arrested at least once. Rehospitalization during follow-up was 51 percent and 42 percent, and the groups did not differ significantly in the total number of days hospitalized during the follow-up period. The participants' perceptions of their quality of life and level of coercion were about the same, and the providers considered patients in the two groups to be similarly adherent to their required treatments (Steadman, Gounis, Dennis, et al. 2001, 333). Some research literature on involuntary outpatient commitment indicates that it appears to improve outcomes in rates of rehospitalization and length of stay. However, studies have serious methodological limitations that are difficult to control (Swartz et al. 1997). OPC may reduce criminal justice contact in persons with severe mental illness. A one-year randomized study of OPC in 262 participants with serious mental illness in North Carolina indicated a significant association with reduced arrest probability (12 percent v. 45 percent) in a subgroup with a prior history of multiple hospitalizations combined with prior arrests and/or violent behavior. Reduction in risk of violent behavior was a significant mediating factor in the association between OPC and arrest. The study found that, in persons with serious mental illness whose history of arrests is related directly to illness relapse, OPC may reduce criminal justice contact by increasing participation in mental health services (Swanson et al. 2001).

The success of this type of legislation depends on adequate funding of community treatment programs. An example of how such a lack of funding can result in problems comes from Ontario, Canada. Underfunding of their involuntary community treatment program has meant that many psychiatric patients who fit the criteria for the program do not receive the designated care. The government did not provide funding for enough support teams, including psychiatrists, who were promised to implement the program under Brian's Law, which was passed in 2000 in memory of an Ottawa sportscaster gunned down by a delusional man. The shortage has created a lengthy waiting list for patients who may be designated for the program, which requires a community treatment order (CTO) that sets conditions for care outside a psychiatric hospital. In Ontario, physicians can obtain CTOs with the consent of either the patient or their designated decision maker. Patients are required to report regularly to their mental health caregivers, and many receive prescription drugs.

Outpatient commitment programs are but one of many programs that add to the costs of the mental health system. Funding remains a major problem in the system.

Funding

The direct costs for diagnosing and treating mental disorders was approximately $69 billion in 1996, and nearly 70 percent of that was for the services of mental health specialty providers, with most of the rest for general medical services providers. Fifty-three percent of mental health costs were paid for by public sector sources, including Medicaid, Medicare, other federal programs, states, and local governments. Almost 60 percent of funding from private funds was from private insurance (U.S. Department of Health and Human Services 1999). Funding problems continued into the 21st century. The National Mental Health Association's State Mental Health Assessment Project found that in 2004 at least twenty-nine states cut funding for mental health services while dealing with budget shortfalls and rising health care costs (National Mental Health Association 2006).

Those seeking funding for care for persons with mental illness may have to compete against other disability groups such as substance abusers and those with developmental disabilities. In some states, such as California, residential care homes for persons with mental retardation receive better financial reimbursements than residential care homes for people with mental illness. Mental retardation may be viewed by legislators more sympathetically, as an accident of nature involving people who are usually not dangerous. Legislators tend to associate mental retardation with children rather than adults, even though most persons with mental retardation are adults. Mental illness is still sometimes seen as volitional, something that people could improve if they would just try, and as we have discussed, mental illness is sometimes associated with violent or strange behavior. Moreover, many people associate mental illness with adults, who are seen as responsible for their actions, even though many children and adolescents are diagnosed as mentally ill.

Medicaid is the core of state public health systems. Medicaid provides an entitlement for medically necessary care for indigent families and dependent children and for children and adults with disabilities. People with severe mental illnesses qualify for Medicaid. The costs of the Medicaid program are shared between the federal government and the states, and there are variations from state to state in additional services the state is willing to pay for. Traditionally Medicaid was a fee-for-service program, under which individuals saw mental health providers who were reimbursed for their fees. However, many providers refused to see people on Medicaid because their fees were not fully reimbursed. Medicaid has been experimenting with mental health services provided through managed care. In California, counties have been entering into memoranda of understanding (MOUs) with health plans that provide physical health care and medications under the Medicaid fee-for-service system. The MOUs establish the criteria for referrals to each other's services. According to these agreements, all mental health services are to be provided by county mental health programs. These programs may deliver the services directly themselves or may refer to services with which they hold contracts.

One of the difficulties in receiving mental health treatment concerns the cost of services and how those costs can be covered. Several approaches have been taken to try to address the cost and payment issues in mental health. The following sections discuss some of those attempts to deal with the problem of funding.

Parity

Over the past two decades, there has been a growing call for parity between mental and physical health care in both the private and public sectors. There has also been a growing body of research calling into question traditional divisions between mental and physical health. Based on that research, many are asserting that the promotion of wellness requires that health care adopt an integrated approach to these two aspects of health. But over the past fifty years, as we have seen, mental and behavioral health services have functioned quite differently from physical health services. Mental health services have been provided unsystematically, have received third-party payment later, and have been stigmatized. Diagnostic and treatment patterns have varied more than in physical health, as no single profession has dominated the field. The field still has wide variability, with a wide array of often competing practitioners who are regulated by different licenses and regulations.

Only 25 percent of persons with a mental disorder receive treatment in the health care system, compared to 60–80 percent of persons with heart disease (National Center for Health Statistics 1992). In 1998, the passage of the Mental Health Parity Act (PL 104-204) helped to increase access to care. The act attempted to bring insurance coverage for mental health services to the same level as services for purely physical disorders. The legislation was limited in reducing insurance coverage discrepancies between physical and mental disorders. However, a number of states have state legislation that requires parity, and 53 percent of the population is covered by those state mental health parity laws (U.S. Department of Health and Human Services 2000, 18–17).

In 1999, California passed a mental health parity law, referred to as Assembly Bill 88 (AB 88). The law requires private health insurance plans to provide equal coverage for physical health and selected mental health conditions, including serious mental illnesses in adults and serious emotional disturbances in children. Typically, health plans have required higher co-payments and deductibles and placed limits on the number of outpatient visits and the number of inpatient days allowed for mental disorders. The law requires health plans to eliminate the benefit limits and reduce the cost-sharing requirements that have traditionally made mental health benefits less comprehensive than physical health benefits.

Failures in private health insurance create significant costs when mental health treatment is shifted to the public sector. AB 88 seeks to decrease the financial burden on California's public sector by ending discriminatory practices in health insurance. Other goals are the expansion of mental health benefits to improve access and quality of mental health services for people with SMI and SED (seriously emotionally disturbed), and the reduction of stigmas associated with mental illness and delivery of mental health services (California Senate Rules Committee 2001).

In the summer of 2001, the California HealthCare Foundation contracted with Mathematica Policy Research (MPR) to conduct an early study of the implementation of the mental health parity law. Through interviews with stakeholders, the research found that, during the first year of implementation, health insurance benefits for mental health services had been expanded in compliance with the law. Also, the law did not appear to have had any adverse consequences on the health insurance market. The stakeholders did identify these remaining challenges:

- The transition to managed behavioral health organizations (MBHOs) by some health plans in response to the law caused initial disruptions in care for some consumers. These disruptions appear to have been exacerbated by inadequate communication efforts and a short lead time for implementing these changes.
- The implementation of "partial parity" for a limited set of SMI and SED diagnoses, rather than all mental health diagnoses, had created administrative challenges and caused confusion for some stakeholders.
- The role of the private sector in delivering services to children with SED needs further clarification, especially given the traditional role of the public sector in providing children's services.
- Consumer education about expanded benefits needs to be improved, in order to facilitate increased access to care under AB 88. (Lake, Sasser, Young, and Quinn 2002, 4)

The researchers concluded that an important goal of AB 88 appeared to have been achieved during the first year of implementation: mental health benefits had been expanded to conform to the parity mandate. But they found it would take additional effort to address the goals of reducing stigma and improving access to care. The law had prompted discussions among stakeholders about additional issues, including the responsibility for additional education efforts, availability of mental health providers in health plan networks, the delivery and management of mental health services under managed care, and the delivery and coordination of services for children by the public and private sectors. They concluded that the full impact of parity might not be known for several years (Lake, Sasser, Young, and Quinn 2002).

Managed Care

In the early 1990s, managed care became increasingly popular. Managed care used a variety of financing mechanisms, such as capitation (paying a set amount per individual) and risk sharing, and oversight mechanisms, such as gatekeeping, which required the primary caregiver to refer to specialists, and utilization review, to reduce health care costs. There were calls to extend managed care to the states' publicly funded systems of mental health service provision (Rochefort 1993; Mechanic 1998). Proponents argued that managed care could improve patient outcomes and increase system efficiency through structural controls and incentives to encourage providers to offer services that were based on clear scientific evidence and medically necessary.

Managed care is having a significant impact on mental health care. Preferred provider organizations (PPOs), health maintenance organizations (HMOs), and point-of-service (POS) plans accounted for 73 percent of the insured population in 1995 (Jensen, Morrisey, Gaffney, and Liston 1997, 126). So many individuals receiving mental health care through their private insurance are under managed care.

A marked development in mental health services has been the development of managed behavioral health care (MBHC) carve-outs, which allows for the separation of health insurance function by disease or service category and allows for contracting separately for those risks. Traditionally, the purchaser, usually an employer, bought all health insurance from a single source. Increasingly, however, purchasers of health insurance are carving out certain benefits to cover under separate contracts at reduced costs, and mental health and substance abuse are increasingly insured through a separate entity.

Throughout the 1990s, states delegated responsibility for provision of mental health services to their clients to private and nonprofit organizations through contracts. As of the end of the 1990s, all but one state Medicaid agency used some managed care arrangements (Donahue 1998). One reason for privatizing these services is to reduce costs. In addition, privatization has been sought to improve the quality and coordination of care and to increase flexibility. Cost savings seem to be concentrated in shifting care from inpatient to outpatient treatment, but evidence on overall use of services showed little change. "The cost control successes of MBHC have provided policy makers and managers with a new and powerful tool. The enormous complexity and diversity of managed care arrangements have made rationing of MH [mental health] care far less amenable to either public or private regulation" (Conti, Frank, and McGuire 2004, 35).

Integration and Linking with Primary Care

Efforts to integrate mental health care with primary (somatic/physical) health care coincide with the development of managed care. One possible solution to the problem of rural mental health care and access to services would be to combine primary care with mental health care. Such a plan might work well in small-town settings, where people can readily identify what services people are going to for assistance, a potential problem in cases where people are unwilling to be identified as going to a mental health clinic. Another option, which is gaining popularity in the southern United States, is providing mental health care through churches. The Southeastern Rural Mental Health Research Center at the University of Virginia found that many rural Southerners, especially African Americans, would be more likely to accept mental health care through their churches (California Healthline 2000).

Integration with primary care has been developed not only in the United States but also in other countries. General practitioners and public mental health services can collaborate to promote continuity of care and early detection of mental disorders. Starting in the 1990s, the Australian government began a campaign to modernize mental health services. The campaign received priority within the national health department. The states and territories began mental health system reforms. One of the objectives was to facilitate collaborative relationships between public mental health services and general practitioners (GPs). The state of Victoria developed a strategy, embodied in their Share the Care Campaign. Guidelines were established for improving collaboration. Here are some of the reasons given for the importance of GPs as mental health providers:

- GPs are often the first person contacted when people begin to experience psychiatric symptoms.

- GPs are a source of support because they often are already familiar to the individual experiencing symptoms and his or her family and loved ones.
- GPs are more accessible for most patients because many insurance agencies require a GP's referral in order to seek specialty mental health services.
- GPs strengthen the continuity of care because they often have a relationship with the individual's family or loved ones and are familiar with the resources of the community.
- GPs provide a high degree of privacy.
- Mainstreaming mental health services into a GP setting normalizes mental health problems within a general health context. (Primary Mental Health Care Australian Resource Center 2001, 33)

Mental Health in the Workplace

One of the ways the chronic problem of underfunding by the government has been met has been through the provision of services by employers. In the past hundred years, employers have offered personal assistance to employees through social welfare (Nelson and Campbell 1990), personnel counseling (Dickson and Roethlisberger 1966; Perrow 1972), and occupational mental health (Carter 1977). World War II brought about recognition of the problems of alcoholism in the workplace, which resulted in the establishment of industrial alcoholism programs (Roman 1988; Sonnenstuhl and Trice 1990).

In 1974, the National Institute on Alcoholism and Alcohol Abuse adopted the term employee assistance program (EAP) to describe job performance–based intervention programs in the workplace. They noted that deterioration in job performance could most often be attributed to the misuse of alcohol, but it could also be related to other problems. Employee assistance programs are employment-based programs for the purpose of identifying troubled employees, motivating them to resolve their problems, and providing access to counseling and other services (Sonnenstuhl and Trice 1990). EAPs have evolved into multiservice programs to address a wide range of problems, including illicit drug use, family problems, mental health problems, and other issues (Roman 1981). An EAP may be an internal administrative unit, or it may be run by an external contractor.

The National Survey of Worksite Health Promotion Activities funded by the U.S. Department of Health and Human Services in 1985 estimated that 24 percent of private, nonagricultural U.S. worksites with fifty or more employees provided an EAP (U.S. Department of Health and Human Services 1987, 23). The Bureau of Labor Statistics in 1988 conducted the Survey of Employer Anti-Drug Programs. They estimated that 6.5 percent of companies of all sizes had an EAP. They estimated 31 percent of employees working in private nonagricultural worksites were covered by an EAP (U.S. Bureau of Labor Statistics 1989, 15). In a follow-up study in 1990, they estimated that 11.8 percent of companies had an EAP. A National Employment Study in 1991 found that 45 percent of full-time employees worked in firms with an EAP (Blum, Martin, and Roman 1992, 219). A later study estimated that 33 percent of private nonagricultural employers with 50 or more full-time employees had an EAP. Approximately 76 percent of employers with more than 1,000 employees had an EAP, while 21 percent of employers with 50 to 99 employees had an EAP. Eighty-one percent of employers' services were provided by external contractors, and 83 percent of services were provided off-site (Hartwell, Steele, French, et al. 1996, 806). Employee assistance programs have continued to grow in the United States and to provide an additional source for mental health services.

When and Why the System Fails

Stigmatization

Clearly, stigmatization of people with mental illness continues to be a problem. This problem can be seen in the resistance of neighborhoods to the establishment of group homes or treatment centers for persons with mental disabilities in neighborhoods, an example of the "not in my backyard" (NIMBY) syndrome. A 1989 Robert Wood Johnson survey showed that proposals to locate mental health facilities in residential neighborhoods were more frequently opposed by civic groups than were shopping malls, homes for AIDS patients, factories, garbage landfills, and prisons (Clark 1994).

Although there are fewer state mental hospitals today than thirty years ago, they continue to play an important role in the mental health care system. In many states, consolidations have occurred, with all mental health services being moved into one or two statewide or regional facilities. In one state that was in the process of consolidating their long-term behavioral health services, a study was done on the attitudes of the community where the hospital was located. Disapproval of consolidation was found not to be related to negative views of mental illness or homelessness; rather it was strongly related to the expected "bad" behavior of the hospitals. The concerns were that the hospital would not prevent escapes and would "dump" former patients onto the community (Wolff and Stuber 2002).

Stigma creates barriers to providing and receiving competent and effective treatment and can lead to inappropriate treatment, homelessness, and unemployment. The elimination of stigma would encourage more individuals to seek needed mental health services. Evidence that mental disorders are legitimate and responsive to appropriate treatment is a potent antidote to stigma.

Housing and Homelessness

People with mental disorders live in a variety of housing situations, including in their own homes, with relatives, in apartments, and in state-regulated adult homes. Many people with mental disorders do not have adequate housing or have no housing and have become part of the population of homeless. Researchers have found that homelessness is much more prevalent among mentally ill individuals than among the general population (Ditton 1999), and homeless mentally ill people often have other serious health problems, including communicable diseases (Essock, Dowden, Constantine, et al. 2003). Homeless mentally ill individuals, particularly men, are at high risk for HIV (Empfield, Cournos, Meyer, et al. 1993; Susser, Valencia, and Conover 1993; Susser, Valencia, Miller, et al. 1995). One study found a 19 percent HIV prevalence among mentally ill shelter users (Susser, Valencia, and Conover 1993, 569). Another study found a 36.7 percent rate of tuberculosis infection. That study suggested that crowded shelter living put HIV-infected mentally ill men at a high risk for tuberculosis (Saez, Valencia, Conover, and Susser 1996).

> Ms. S., living in a shelter in Washington, D.C., was hospitalized in 1984 at St. Elizabeth's Hospital because of her clearly psychotic behavior (crawling on the street on her hands and knees, running in front of cars, eating

> eggs shell and all). She was diagnosed with schizophrenia and her condition improved modestly with antipsychotic medication over the next eight months. Either because of her illness or her personality traits she was not fully cooperative with hospital routines, often refusing to attend scheduled activities.
>
> One day she was told by the staff that if she refused to go to occupational therapy she would be discharged. She did refuse, and was discharged the same day. She was escorted from the ward wearing a cotton dress, thin sweater, and sneakers, but no underclothing or socks; the temperature was projected to go below freezing that night. She was given a list of shelters, a week's supply of medicine, and told to make her own appointment for aftercare. The next day, when I evaluated her in a shelter, she was still markedly psychotic. (Torrey, Bargmann, and Wolfe 1985, 8)

Deinstitutionalization, which affected approximately 2 million seriously mentally ill in the United States, led to at least twice as many seriously mentally ill people living on streets and in shelters as in public hospitals. The public gradually recognized the problem of the homeless in the early 1980s. Approximately one-third of homeless individuals are seriously mentally ill (Torrey 1988). Estimates vary by city. In a 1986 survey conducted by the U.S. Conference of Mayors, the percentage of people with mental illness, excluding alcohol and drug abuse, in the homeless population ranged from 60 percent in Louisville to 15 percent in Philadelphia, Los Angeles, and New Orleans (U.S. Conference of Mayors 1986, 44). Three studies of people with schizophrenia in homeless shelters during the 1980s found that between 36 percent and 39 percent of the total shelter population had schizophrenia in Washington, Boston, and Philadelphia (Arce et al. 1983; Bassuk, Rubin, and Lauriat 1984; Torrey et al. 1985). A 1988 homeless study in Los Angeles found that 44 percent of the homeless surveyed had had a psychiatric hospitalization (Gelberg, Linn, and Leake 1988). Another 1988 study, this one done in Ohio, found that 36 percent of patients discharged from state mental hospitals were homeless within six months (Belcher 1988). In 1989, a Massachusetts study found that 27 percent of discharged psychiatric patients were occasionally or predominantly homeless within a six-month period (Drake, Wallach, and Hoffman 1989).

> I first saw Ms. Z in a women's shelter in Washington. She was extremely paranoid and delusional but acknowledged that she had been married to a professional man and had raised several children. She refused to take medication. She left the shelter, and I next heard that she had been living on the mezzanine of National Airport for several months; it was apparently the only place she felt safe from Israeli secret agents who she believed were injecting her in her sleep and were responsible for her voices. One of her daughters, on her way back from college, accidentally discovered her mother at the airport. (Torrey 1988, 10)

As discussed in Chapter 1, those who have left mental health hospitals during the process of deinstitutionalization and who have been able to find housing have often found inadequate housing. In New York, approximately 15,000 adults with mental illness, including many low-income members of minorities, reside in more than 100 adult homes across the state. For thirty years, state investigators have characterized the homes as "little more than psychiatric flophouses, with negligent supervision and incompetent distribution of crucial medication" (Levy 2002). While some homes try to provide good care, all have the same systemic problems, such as unskilled workers and gaps in resident supervision. Most of the operators of the homes have no mental health training and are operating commercial facilities where beds must be kept full to make money. This results in some operators keeping residents in the facility instead of referring them to a hospital or accepting residents who are a threat to themselves or others (Levy 2002).

Families are the fastest-growing element of the homeless population (Anderson and Imle 2001; Rog, Holupka, and McCombs-Thornton 1995). Families make up one-third of the homeless population (Wright, Valdez, Hayashi, et al. 1990, 230), and female-headed families make up between 70 percent and 90 percent of homeless families (Anderson and Imle 2001, 400; Bassuk 1990, 1050). Women make up about 20 percent of the adult homeless population, and homeless women without relatives have been found to be more psychiatrically disabled than other subgroups of the homeless (Burt and Cohen 1989, 5), especially women in mid-adulthood and older (Rossi, Fisher, and Willis 1986, 6). Unsurprisingly, mentally ill women who are homeless are thought to need special services (Bachrach 1987).

Studies have found that the homeless or residentially unstable have greater alcohol and or drug abuse problems (Drake, Wallach, and Hoffman 1989; Drake, Wallach, Teague, et al. 1991; Susser, Lin, and Conover 1991). Studies have also found higher symptom levels and greater noncompliance with prescribed treatments among this population (Drake, Wallach, Teague, et al. 1991; Susser, Lin, and Conover 1991). A study of risk factors for homelessness found that homeless women with schizophrenia were different from never-homeless counterparts, in that a higher proportion of them had a concurrent alcohol abuse diagnosis and a significantly greater number had a concurrent diagnosis of antisocial personality disorder. In addition, family support was less adequate for the homeless. They found that poor family support was a risk factor for the persistence of homelessness. However, adequacy of family support was a more important risk factor for women than the other factors (Caton, Shrout, Dominguez, et al. 1995). A study of schizophrenic men also found the homeless to be more likely to have concurrent alcohol and/or drug abuse, antisocial personality disorder, and poor family support (Caton, Shrout, Eagle, et al. 1994).

Favorable outcomes have been found for the provision of services such as treatment for substance abuse and psychiatric disorders to the homeless population (Pollio, Spitznagel, North, et al. 2000). The characteristics of systems of care and organizations may play an important role in service delivery to homeless substance abusing populations (Sosin 2001). Fragmented service planning for the homeless has been shown to impede the effective delivery of a continuum of services for mental disorders and substance abuse (Calloway and Morrissey 1998; Randolph, Blasinsky, Morrissey, et al. 2002).

A study of twenty-three organizations in the St. Louis area providing services with a homelessness, mental health, or substance abuse focus found that mental health service use was associated with funding simplicity, professionalism, and organizational size. Substance abuse service use was associated with funding diversity, professionalism, and focus of services on substance abuse service provision. Shelter service use was associated with diversity of services. The study also found a low rate of service use of mental health and substance abuse services in relation to the rates of substance abuse and mental health problems in the population (North, Pollio, Perron, et al. 2005).

The Stewart B. McKinney Homeless Assistance Act was passed by Congress in 1987 and made available $180 million in housing assistance for homeless people. The McKinney Act requires state plans coordinating food and shelter, health, education, and job training services. There is also an Interagency Council on the Homeless in the executive branch, which coordinates the homeless assistance programs of various federal agencies. The Interagency Council on the Homeless requested the National Institute of Mental Health to review the problem of homelessness and mental illness. An interdepartmental task force chaired by NIMH released a report in 1992 that indicated fifty steps the federal government should take and additional actions to be taken by state and local governments and the private sector (Federal Task Force on Homelessness and Severe Mental Illness 1992). The report set forth four goals: integration of needed health, mental health, social, and housing services; increased housing options and alternative residential services; linkage of homeless individuals with existing benefit programs for which they qualify; and distribution of information on homelessness and service innovations. In May 1993 President Clinton signed an executive order directing the seventeen federal agencies that make up the Interagency Council on the Homeless (ICH) to prepare a coordinated federal plan for breaking the cycle of existing homelessness and for preventing future homelessness. Efforts have continued by the federal, state, and local governments to provide housing and services to the homeless mentally ill.

Opponents of deinstitutionalization have used the issue of homeless mentally ill to discredit mental health practices and argue that deinstitutionalization causes homelessness (Appelbaum 1987; Isaac and Armat 1990; Wyatt and DeRenzo 1986). Leona Bachrach (1992) argued that looking at deinstitutionalization in broad terms as state hospital depopulation, diversion of admission to other social and medical settings, and a civil rights atmosphere that discourages some distressed persons from entering the mental health system suggested that deinstitutionalization has a fundamental relationship to homelessness. Others believe that homelessness comes from poverty, a diminishing supply of low-cost housing, and the inability of low-income persons to afford available housing (Rossi 1989; Rossi and Wright 1987). A study published in *American Psychologist* (Aiken, Somers, and Shore 1986) argued that many mental health officials believe housing is their most serious program shortcoming and report that only a fraction of appropriate housing is available for their patients. "Lack of suitable housing contributes to episodic hospitalization and remains a major barrier to releasing mental hospital patients who are judged legally and clinically ready for discharge" (Rochefort 1993).

Those with co-occurring disorders are most prone to be in the homeless population. Co-occurring mental and addictive disorders are estimated to affect 50 to 60 percent of homeless persons (Interagency Council on the Homeless 1992). Living on the streets also exposes the mentally ill to more opportunity to come in conflict with the law, which can lead to their incarceration in the criminal justice system, as the next section makes clear.

Mental Illness and Criminal Justice

Since the country began, mentally ill people have been incarcerated. But with the development of psychiatric hospitals, awareness grew of the existence of a group of mentally ill individuals who were both mentally ill and criminally dangerous. A number of hybrid arrangements were created in both prisons and psychiatric hospitals to address this issue (Morris 1995). In prisons, the criminally mentally ill were housed in the general population or in special psychiatric units. In some large states, separate prison psychiatric hospitals were established. By 1926, there were more than 100 psychiatrists and psychologists working in correctional settings (Rotman 1995). State psychiatric hospitals developed secure wards for mentally disordered criminal offenders. Minimal treatment was provided, along with security similar to that in prisons. This divided system has continued in many states in the twenty-first century (Fagan 2003).

The number of mentally ill in the correctional system increased sharply at approximately the same time as the passage of the Mental Retardation Facilities and Community Mental Health Centers Act of 1963. With a lack of community resources, the mentally ill were at risk for all sorts of problems in daily living, as discussed above. Many individuals were unable to obtain employment and became homeless. "Poverty and accompanying mood instability set the stage for criminal behavior such as theft, vandalism, trespassing, and the use of drugs and alcohol. After passage of this act, many mentally ill individuals who would otherwise have been referred for mental health care after causing disturbances or engaging in criminal acts were instead remanded to the criminal justice system" (Holton 2003, 102). Torrey (1995) observed that prisons and jails had become America's new mental hospitals.

During the twentieth century, the medical model emphasizing diagnosis and treatment was primary. Rehabilitation programs were used, but by the 1970s, pessimism was growing, as rehabilitation seemed to have little impact on the recidivism rate of treated offenders (DiIulio 1991; Martinson 1974). By the late 1980s, however, the focus was still on correctional programs and rehabilitation. These programs included education and vocational training, as well as mental health, work, and religious programs. The programs were designed to develop skills that could be used upon release (Fagan 2003).

Crime, criminal justice costs, and property loss associated with mental illness costs $6 billion a year, and people with mental illnesses are overrepresented in jail populations (U.S. Department of Health and Human Services 1999). Many of these inmates do not receive treatment for mental illness. A recent report by the Pacific Research Institute for Public Policy on criminal justice and the mentally ill noted that the total costs for state and local governments for arrest, processing in court, and jail maintenance of people with mental illnesses exceeds the total state and local government expenditures on mental health care (California Council of Community Mental Health Agencies n.d.).

> Wayne B., a 31-year-old man who has been afflicted with paranoid schizophrenia since age 17, responds well to antipsychotic medication. He was under my care for four years and, on medication, was able to share an apartment with three other released patients and hold a half-time job. Then his outpatient care was transferred to a local community mental health center, which simply dropped him from its rolls when Mr. B. did not keep an appointment. He ran out of medication and was lost to follow-up until three months later when he was arrested for stabbing a neighbor who he thought was trying to hurt him. Mr. B. is now serving a long sentence in the District of Columbia jail. (Torrey 1988)

In 1998, in two studies conducted by the Bureau of Justice Statistics (Beck and Maruschak 2001; Ditton 1999), there were approximately 284,000 to 291,000 offenders with mental illness incarcerated in the nation's jails and prisons. According to a 1999 Justice Department report, more than 16 percent of adults in jails and prisons nationwide have a mental illness, and more than 20 percent of those in the juvenile justice system have serious mental health problems (Pyle 2002). A national study by Paula Ditton found that 16 percent of all probationers, 16 percent of all local jail prisoners, 16 percent of all state prisoners, and 7 percent of all federal prisoners had a mental condition. She found that state prisoners with mental conditions were more likely to be in prison for violent offenses (53 percent v. 46 percent). They were more likely to be under the influence of alcohol or drugs at the time of their arrest (59 percent v. 51 percent), and they were more than twice as likely as other prisoners to have been homeless during the twelve months prior to their incarceration (20 percent v. 9 percent) (Ditton 1999, 1–2). There are more mentally ill people in U.S. prisons and jails (283,000 in 1998) than in mental hospitals (61,772 in 1996) (Parker 2001).

Of this population, many are in city or county jails. Such jails house people awaiting court and may incarcerate those serving terms of less than one year. The study by Ditton referred to above found that approximately 19 percent of all jail and prison inmates had severe mental illnesses. Similar findings have been reported in other studies of jails (Diehl and Hiland 2003; Diehl and Porter 2004; Roy and Ruddell 2004). Rick Ruddell, in his study of mentally ill populations in jails, found that the mean estimate of persons with mental illness was 13.1 percent (Ruddell 2006, 2). He also found that the only variable that was significantly associated with high or low percentages of jail inmates with mental illness was the size of the facility. Larger facilities reported higher percentages of mentally ill inmates. Almost 80 percent of his survey participants believed that admissions of persons with mental illness in their jails had increased during the past five years, and 70.5 percent of administrators and mental health specialists felt that their jails were responding to more inmates with co-occurring disorders.

Most people spend a short time, often a day or two, in jail, before being released on bail, but G. Larry Mays and Rick Ruddell (2004) found that a subpopulation of inmates is long term, and this subpopulation includes people with mental illness, some of whom may spend months or years awaiting the resolution of their court case. Other people with mental illness have multiple admissions over a period of years, usually for minor offenses.

They are sometimes called frequent fliers (Ford 2005). Mentally ill persons in jail have a higher potential for causing disruptions (Roy and Ruddell 2004). Ruddell (2006) found that, of four populations examined (persons with mental illness, frequent fliers, gang members, and inmates who had served at least one year in jail), persons with mental illness had the highest involvement in problem behaviors. They were perceived to be involved in problem behaviors at nearly twice the rate of the next nearest group, gang members. Inmates who were mentally ill were believed to be the group most likely to assault staff members, and they were also believed to be the group most likely to be assaulted by other inmates. Human Rights Watch (2003) has also found them to be at higher risk of being victimized.

On any given day, about 3,600 people with severe mental illnesses are in the Los Angeles County jail, 700 more than in the country's largest mental hospital. A 1992 survey of 1,400 jails in all 50 states found substantial numbers of mentally ill inmates, nearly a third of them confined without criminal charges (Torrey 1992).

Researchers have found that individuals with serious mental illness are more likely than non-disordered offenders to be arrested more than once for the same crimes (Kilborn 1999; Teplin 1984). Also, in some places homelessness and related crimes, such as sleeping in public places or panhandling, are against the law (Bumiller 1999; Herszenhorn 1999). The courts are also more likely to incarcerate the mentally ill because of the public's beliefs about the potential for violence from the seriously mentally ill. Researchers found an unrealistically elevated fear of violence from mentally ill persons and a strong desire to maintain social distance from them (Link, Phelan, Bresnahan, et al. 1999). Approximately 50 percent of people who are diagnosed with both severe mental illness and substance abuse problems have had previous contact with the criminal justice system (Theriot and Segal 2005). One study (Wolff, Diamond, and Helminiak 1997) found that, among clients of an assertive community treatment (ACT) program, approximately half of the criminal justice contacts could be classified as related to a disturbance rather than a more serious offense. More than one-third of the disturbance calls were for police protection of clients with mental illness who were the victims of a crime.

Assertive community treatment (ACT) is an evidence-based model for providing services to people with severe mental illness which has been found to result in better clinical outcomes for people with severe mental illness who have many treatment failures and prior hospitalizations (Bond, Drake, Mueser, and Latimer 2001). ACT has also been found effective with those who are homeless (Morse, Calsyn, Allen, et al. 1992; Morse, Calsyn, Klinkenberg, et al. 1997). However, it is not more effective than other treatments in reducing contact with the criminal justice system (Bond, Drake, Mueser, and Latimer 2001; Mueser, Bond, Drake, and Resnick 1998). The ACT approach involves services provided by a multidisciplinary team, with a staff-to-client ratio of one to ten; most services provided in a community setting; no time limits on services; daily discussion of clients by team members; and availability of a team member twenty-four hours a day.

Thomas Fagan (2003) argued that three factors have contributed to the rapid increase in the number of persons with mental disorders in the criminal justice system. First, the law-and-order mentality that began during the 1980s led to tougher sentencing legislation, in some cases abolished parole, changed competency/criminal responsibility standards, and added three strikes laws. Now more people are in prison for longer periods of time, and as the number of people in prison increases, so does the number of people in prison with mental disorders. Second, as mental hospitals have downsized and mental patients have been returned to the community without adequate services, more people with mental disorders have appeared in the homeless and prison populations (Torrey 1988; Torrey, Stieber, and Ezekiel 1992). Many of the functions of state psychiatric hospitals have been assumed by the criminal justice system (McConville 1995). Third, drug enforcement has increased the number of people imprisoned for drug offenses. In 1980, 6 percent of the prison population was incarcerated for drug offenses and in 1996, 23 percent of the prison population was imprisoned for drug offenses (Begun, Jacobs, and Quiram 1999). Sixty-three percent of the federal prison population was in prison for drug offenses (Mumola 1999). In 1998, only 14 percent of incarcerated people were in drug treatment (Camp and Camp 1998, 52). Criminal history variables are the best predictors of recidivism for both mentally disordered offenders and non-disordered offenders. Substance abuse and serious mental illness are modest predictors of recidivism (Bonta, Law, and Hanson 1998).

Many prisoners in the criminal justice system have severe psychiatric disorders and require mental health services (Guy,

Platt, Zwerling, and Bullock 1985; Lamb and Grant 1982; Teplin, Abram, and McClelland 1996). Women in jail have higher rates of severe mental disorder than men (Teplin, Abram, and McClelland 1996). This is particularly true of depression (Teplin 1990b and Teplin 1994).

Court decisions have established that mentally ill prisoners have a constitutional right to treatment (*Estelle v. Gamble,* 429 U.S. 97 [1976]; *Bell v. Wolfish,* 441 U.S. 535, n. 16, 545 [1979]). Court decisions have also determined that they should receive services on an emergency and regular basis (*Bowring v. Godwin,* 551 F.2d 44 [4th Cir. 1977]; *Inmates of Allegheny County Jail v. Pierce,* 487 F. Supp. 638, 642–643 [W.D. Pa. 1980]; *Finney v. Huto,* 410 F. Supp. 1026, 1037 [E.D. Ark 1976]; *Jones v. Wittenberg,* 323 F. Supp. 793 [N.D. Ohio 1971]; *Finney v. Mabry,* 534 F. Supp. 1026, 1037 [E.D. Ark 1982]). Under the law inmates have a right to mental health services, just as they have a right to medical treatment (*Ramos v. Lamm,* 639 F.2d 559, 577 [10th Cir. 1980]; *Hoptowit v. Ray,* 683 F.2d 1237, 1253 [9th Cir. 1982]).

It has been argued that diversion may be the only alternative to the cycling of the mentally ill in and out of the criminal justice and forensic mental health systems (Rogers and Bagby 1992). Torrey and his coauthors recommended that jail diversion programs be set up to minimize the number of mentally ill persons in jails (Torrey, Stieber, and Ezekiel 1992). In 1991, a National Coalition for the Mentally Ill in the Criminal Justice System report also recommended diversion of the mentally ill into appropriate mental health services. Henry Steadman, Sharon Barbera, and Deborah Dennis defined diversion programs as "specific programs that screen defined groups of detainees for the presence of a mental disorder; use mental health professionals to evaluate those detainees identified in screening; negotiate with prosecutors, defense attorneys, community-based mental health providers, and the courts to produce a mental health disposition as a condition of bond, in lieu of prosecution, or as a condition of a reduction in charges (whether or not a formal conviction occurs); and link the detainee directly to community-based services" (Steadman, Barbera, and Dennis 1994, 1112). A growing number of state mental health agencies are collaborating with state and local corrections agencies to deliver mental health services in jails and prisons. They are also working on programs to divert people with mental illness from the criminal justice system. Pre-jail diversionary programs refer minor offenders to community-based intervention programs rather than placing them in jail. Those who are accused may also be found unfit to stand trial or not guilty by reason of insanity. Women are disproportionately represented in both these categories (Steadman, Monahan, Hartstone, et al. 1982). People found unfit to stand trial or not guilty by reason of insanity may be placed involuntarily in a mental health facility until they are found fit, either to stand trial or to be returned to the community. There are also postbooking diversion programs, after the person has been processed into jail, providing placement in community mental health programs upon release from jail. Approximately 40 percent of dual-diagnosed people charged with a criminal offense report that the judicial system tries to induce them to receive mental health treatment by offering to reduce criminal charges or jail time, a practice referred to as leverage (Monahan, Redlich, Swanson, et al. 2005).

There are two main types of diversion programs. The first is police-based and affects offenders before they are booked. Postbooking programs are administered by the court or the jail authorities, and there are three subtypes: prearraignment, postarraignment, and mixed (Steadman, Morris, and Dennis 1995). Henry Steadman, S. M. Morris and Deborah Dennis found six factors that were consistently found in the most effective postbooking jail-diversion programs: "(1) integrated services, (2) regular meetings of key agency representatives, (3) boundary spanners, (4) strong leadership, (5) early identification, and (6) distinctive case management services" (Steadman, Morris, and Dennis 1995, 1631). Discharge planning and follow-up they found to be critical aspects of a successful diversion program. However, few programs visited by Steadman, Morris, and Dennis had follow-up procedures for diverted detainees, and even when there was linkage to community-based services, few programs had mechanisms to maintain the initial linkage. Some programs did work, through case management, outpatient services, and residential placement, to ensure that when detainees left the jail there was a place for them. Those transitional programs also helped clients to locate permanent housing, find suitable employment, and develop life skills. They asserted, "The most effective diversion programs are those that are part of a comprehensive array of other jail services, including screening, evaluation, short-term treatment, and discharge planning (i.e., linkage), that are integrated with community-based mental health, substance abuse, and housing services" (1634).

Despite the law, many jail detainees do not receive needed services (Morgan 1981; Steadman 1992; Steadman, McCarty, and Morrissey 1989; Whitmer 1980). Linda Teplin (1990b) found that 37 percent of male jail detainees who had schizophrenia or major affective disorder received services while in jail (Teplin 1990b, 291). George Camp and Camille Camp in an annual survey of federal and state correctional facilities found the percentage of offenders who were involved in mental health programs ranged from a low of less than 1 percent in several smaller states to a high of 18 percent in Ohio (Camp and Camp 1998, 10). Ruddell (2006) found that less than 40 percent of jails had mental health units, even though the average size of the jails in his sample approached 1,000 beds. Litigation has forced some jails to provide better care to inmates needing mental health services (Schlanger 2003). Women are relatively underserved (Velimesis 1981). A 1991 to 1993 study of 1,272 arrested females awaiting trial in the Cook County Department of Corrections in Chicago, Illinois, found of the 116 women needing mental health services, only 23.5 percent received those services during their jail stay (Teplin, Abram, and McClelland 1997, 605–606). This compares to 35.5 percent of men needing mental health services and receiving services in 1983–1984 (Teplin 1990b). The women's diagnoses determined whether or not they received mental health services. The most services were provided for schizophrenia or manic episodes (47.5 percent). Only 15.2 percent of detainees were treated for depression (Teplin, Abram, and McClelland 1996, 607). This low treatment rate was in spite of the fact that 36 percent of all jail inmate deaths nationally are the result of suicide (U. S. Department of Justice 1995).

Inmates with a treatment history were more likely to receive services. Of detainees with schizophrenia or manic episode and a treatment history, 75 percent received mental health services, compared with only 27 percent of inmates who had these disorders but no history. Only 3 percent of detainees with a major depressive episode but no treatment history received services. But detainees with schizophrenia or manic episode and a co-occurring drug use disorder were less likely to receive services than those with only the mental health disorder. Prisoners with only one prior arrest or no arrest history were more likely to receive services than those with two or more arrests (Teplin, Abram, and McClelland 1996, 607).

Another approach established in the early 1990s is mental health courts. These specialized courts hear primarily nonviolent misdemeanor cases. These courts have a specialized docket of defendants who have mental health problems, and they have a diversionary orientation. They are able to use the power of the criminal justice system to get compliance with treatment requirements in exchange for release or probation (Griffin, Steadman, and Petrila 2002). Success is tied to the availability of court or probation staff to monitor the defendant's progress (Steadman, Davidson, and Brown 2001). One evaluation suggests that these courts are effective at increasing access to mental health treatment (Boothroyd, Poythress, McGaha, and Petrila 2003). Probation departments in some cases have established specialized caseloads for offenders with mental disorders, and comprehensive discharge planning, including specialized probation units, has been successful in some jurisdictions (Wolff, Plemmons, Veysey, and Brandli 2002). Also the use of case management plans in jail increases the chances that an offender will continue to participate in community services (Ventura, Cassel, Jacoby, and Huang 1998). Nevertheless, the effectiveness of mental health courts is not yet fully proven (Trupin and Richards 2003).

T. J. Fagan (2003) described a three-level model for provision of mental health services in correctional facilities. Level 1 provides mental health services to individual prisoners. Accrediting bodies such as the American Correctional Association and the National Commission on Correctional Health Care often mandate these services. These services are for detection, diagnosis, both short- and long-term treatment, and referral of prisoners with significant mental health problems. Services provided include initial intake assessment, crisis intervention, brief counseling, individual psychotherapy, case management, segregation reviews, and special mental health evaluations. Initial intake assessment may identify emotional, intellectual, and behavioral deficits or significant mental impairment. The assessment may also identify specialized treatment needs, including drug or alcohol detoxification or treatment, suicide monitoring, and psychotropic medication. Information may also be provided to other prison staff about special housing or program needs, potential adjustment problems, assault potential, and escape risk.

Level 2 services provide programs and services to specific groups who have similar problems or characteristics or who share certain mental health treatment needs. There is strong evidence that some correctional group treatment programs are effective at reducing recidivism rates (Gendreau 1996; Gendreau, Goggins, and Smith 1999). Fagan has listed a number of elements derived from studies published on the subject that are necessary for effective correctional treatment. These include:

1. Programs are presented in a correctional environment that is both supportive of programming initiatives and stable in terms of social climate and staff resources.
2. Stakeholders in the program (e.g., community sources, correctional administrators, program providers, correctional staff) agree that the program has value and is consistent with the institution's mission and values.
3. Treatment providers are appropriately trained and have professional credibility with offenders, staff, and correctional administrators.
4. Role models are provided to offenders that are within their reach (e.g., corrections employees or community figures as opposed to sports celebrities).
5. Programmatic goals are clearly stated and target specific factors known to lead to crime, such as pro-criminal attitudes and associations, impulsivity, and risk-taking behaviors.
6. Programmatic emphasis is on basic cognitive-behavioral and problem-solving skills that clearly relate actions to consequences and demonstrate respect for authority.
7. Relevant postprogram support systems are developed that use whatever human and financial resources are available within both correctional and community settings to aid and support program goals. (Fagan 2003, 12–13)

One of the most common Level 2 services is a drug abuse treatment program. Such programs in jails and prisons use many of the same treatment models that are used in the community, including Alcoholics Anonymous and Narcotics Anonymous. Assessment of several substance abuse programs in prisons and community follow-up settings found positive after-release outcomes for treatment participants compared to prisoners who did not receive treatment (Lipton 1995).

Level 3 services involve consultative services, where mental health professionals assist correctional administrators. For example, mental health professionals have developed classification systems to separate predatory offenders from less threatening offenders. Also mental health professionals have used their conflict resolution skills to mediate disputes between facility departments and between management and labor.

Ruddell (2006) has found a three-pronged strategy to be the best way to deal with persons with mental illness in jails: community and jail-based diversionary programs; comprehensive mental health programs within jails, including mental health units and case managers; and criminal justice involvement with broad-based community programs for persons with mental illness.

A trend to create supermax correctional facilities—correctional units characterized by solitary housing, including exercising and eating alone—is also having an impact on the incarceration of the mentally ill. Inmates in those units are provided limited treatment and other programs. Researchers have found that mentally ill inmates are overrepresented in these isolation units (Harrington 1997; Jemelka, Trupin, and Chiles 1989; Toch and Adams 1986). Breaking of prison rules may occur because of an inmate's mental illness, and rule breaking may result in the prisoner being placed in a supermax unit. Researchers have also found that the extreme isolation and deprivation of supermax units can contribute to mental illness (Grassian 1983; Haney and Lynch 1997; Harrington 1997). Some systems screen and divert mentally ill offenders from supermax units, but not all (Harrington 1997).

One aspect of the criminal justice system that has become especially problematic for the mentally ill has been a new approach to handling sexually violent predators. Over the past ten to fifteen years, a number of state legislatures have become aware of the risk presented by sexually violent predators (Feldman 1997; Freeman-Longo and Knopp 1992). In 1990, the state of Washington enacted the first law focused on sexually violent predators, and that law was upheld by the U.S. Supreme Court in 2001 in *Seling v. Young* (99–1185 19 2 F. 3d 870 Ct. 531 U.S. 250 [2001]). A number of other states have followed suit. The laws differ, but they generally provide for the civil commitment and long-term care of designated sexual offenders when they are released from prison (Feldman 1997; Teir and Coy 1997). The Kansas law was the first such law to be challenged and reviewed by the Supreme Court. In 1997, in *Kansas v. Hendricks* (117 S. Ct. 2072), the Supreme Court upheld Kansas's Sexually Violent Predator Act of 1995. The Kansas law provides for postsentence, involuntary civil commitment of individuals who have a mental abnormality or personality disorder that makes them more likely to engage in predatory acts of sexual violence. The act defines a mental abnormality as a "congenital or acquired condition affecting the emotional or volitional capacity which predisposes the person to commit sexually violent offenses in a degree constituting such person a menace to the health and safety of others"(*Kansas v. Hendricks,* 117 S. Ct. 2072, 13).

The act requires the state agency with authority over the release of a potential sexually violent predator to notify the local prosecutor. The prosecutor may file a petition invoking the involuntary civil commitment procedure. A federal district court then determines probable cause to conduct a trial. The trial decides whether the person is a sexually violent predator. If a person is committed under the act, the committing court must hold an annual review of the mental condition of the sex offender to decide whether detention should be continued. The civilly committed person may petition for release at any time.

The number of sex offenders in state prisons quadrupled from 1980 to 1994, with 88,100 offenders incarcerated in 1994 (National Center for Missing and Exploited Children Sex Offender Policy Task Force 1998). There is no currently universally accepted treatment approach for treating incarcerated sex offenders (Boardman and DeMatteo 2003). The most common kind of treatment program uses some form of cognitive therapy, along with psychopharmacological or behavioral techniques (State of Minnesota 1994). Treatment outcome studies report widely differing conclusions regarding the effectiveness of treatment (Boardman and DeMatteo 2003). Research suggests that some treatment is effective. G. C. N. Hall found a recidivism rate of 19 percent in treated offenders and 27 percent in a comparison group of nontreated offenders (Hall 1995, 804). He concluded that cognitive-behavioral and pharmacological treatments were more effective than just behavioral approaches. Another study found that cognitive-behavioral programs are the most effective treatment for child molesters (Marshall, Jones, Ward, et al. 1991).

Grading the States 2006

Is more progress being made in providing effective mental health services during the twenty-first century than during the twentieth century? The National Alliance for the Mentally Ill (NAMI) continued their evaluation of state mental health services by issuing a new report, *Grading the States 2006: Report on America's Health Care System for Serious Mental Illness*. They recognized that each state mental health authority (SMHA) has responsibility for administering all mental health services in that state, even though other agencies than mental health agencies, such as Medicaid and agencies involved in vocational rehabilitation, housing, and corrections, are involved in providing mental health services. NAMI defined ten elements as characterizing a high-quality state mental health system:

1. Comprehensive services and support
2. Integrated systems
3. Sufficient funding
4. Consumer- and family-driven systems
5. Safe and respectful treatment environments
6. Accessible information for consumers and family members
7. Access to acute care and long-term care treatment
8. Cultural competence
9. Health promotion and mortality reduction
10. Adequate mental health workforce

Comprehensive services and supports include affordable and supportive housing, access to medications, assertive community treatment, integrated dual-diagnosis treatment, illness management and recovery, family psychoeducation, supported employment, jail diversion, peer services and supports, and crisis intervention services. Funding through Medicaid, the Federal Mental Health Services Block Grant, or the states should achieve reduced symptoms, increased independence, employment, housing, and increased consumer satisfaction. "States also should be able to show that these expenditures reduce negative outcomes such as hospitalizations, homelessness, criminal justice involvement, and suicides" (NAMI 2006, 199).

Their report does not suggest any improvement in the twenty-first century. Rather, most states have either kept their funding level the same or have reduced funding for mental health services, leaving large numbers of people without access to services and increasing the number of people with mental illness in emergency rooms and jails.

The 2006 NAMI evaluation developed a report card for each state by assessing each state's mental health system relative to three documents: the U.S. Surgeon General's 1999 *Report on Mental Health,* the president's New Freedom Commission on Mental Health 2003 book *Achieving the Promise: Transforming Mental Healthcare in America,* and the Institute of Medicine of the National Academy of Sciences 2005 book *Improving the Quality of Health Care for Mental and Substance Abuse Conditions.* The evaluation used five evidence-based practices promoted by the Center for Mental Health Services along with other treatment- and recovery-oriented services. Each state was rated with a letter grade from A to F based on four sources (State Mental Health Authority Self-Reported Questionnaire, public information, a Consumer and Family Test Drive, and interviews) and scored from thirty-nine specific criteria in four categories: infrastructure, information access, services, and recovery supports.

The United States scored a D on the national report card. "For too many Americans with mental illnesses, the mental health services and supports they need remain fragmented, disconnected and often inadequate, frustrating the opportunity for recovery. Today's mental health care system is a patchwork relic—the result of disjointed reforms and policies. Instead of ready access to quality care, the system presents barriers that all too often add to the burden of mental illnesses for individuals, their families, and our communities" (NAMI 2006, 31).

Insufficient information was received from Colorado and New York, and they were not graded. Of the remaining states, no state received an overall grade of A. Two states, Connecticut and Oregon, received A grades on infrastructure, while 10 states received a grade of F. Three states received a grade of A for information access: Ohio, South Carolina, and Tennessee, while seventeen states received a grade of F. No state received a grade of A for services: the best grade was B+, for Wisconsin, and three states, Illinois, Kentucky, and Montana, received an F. Five states, California, Hawaii, Maine, Minnesota, and Vermont, received an A for recovery supports, while seven states received a grade of F.

California is the most populous state in the country, with more than 35 million people. Overall, California received a grade of C from NAMI. The category grades were B for infrastructure, C- for information access, D for services, and A for recovery supports. California was fourteenth in per capita spending at $109.34, compared to number 1 for Washington, D.C., at $414.08, number 2 for Pennsylvania at $195.01, and number 51 for New Mexico at $28.80. Per capita income in California was ranked tenth at $32,043, compared to the District of Columbia at number 1 with $45,898, Connecticut at number 2 with $40,990, and Mississippi at number 51 with $22,263. California spent the most in total mental health spending at $3,862 million, while Idaho spent the least at $46 million. The suicide rank in California was 42 in a tie with Delaware while the number 1 state ranked for highest suicide rate was Wyoming, and the number 51 ranking was Washington, D.C. (NAMI 2006, 208).

There are signs of progress in California, as there are in other states. The passage of Proposition 63, Mental Health Services Act, which will raise taxes on millionaires, has led to county planning on how to use the projected $750 million to be raised by the tax. (California provides mental health services through its fifty-eight counties and two city agencies.) Emphasis is being placed on services for the unserved and underserved. "In 2003, the California Mental Health Planning Council estimate that approximately 300,000 adults with serious mental illnesses did not have access to needed services. . . . Many indigent individuals in the state go without desperately needed mental health services simply because they don't exist" (NAMI 2006, 45).

California is moving in the direction of evidence-based practices (EBPs), and the state is piloting the federal Assertive Community Treatment (ACT) model in four counties. In 2004 a law to reduce the use of seclusion and restraints was passed. However, the state hospitals remain a major problem. Issues of abuse and neglect arose recently at Napa State Hospital in the northern part of the state and Metropolitan State Hospital in Los Angeles. To address persons with mental illness in the criminal justice system, thirteen local mental health courts have been established. To improve employment of people with mental disabilities, programs for supported employment have been developed in 22 counties (NAMI 2006, 46).

Thus when NAMI (2006) identified a sample of state innovations and best practices in the states, in financing they recognized the passage of Proposition 63 in California. They also recognized Arizona and Colorado for allowing local municipalities to raise funding for mental health through special tax districts or bonds. In housing improvements, NAMI recognized Tennessee's effort in developing housing, which had risen from almost nonexistent in the 1990 NAMI report to among the best. They also recognized Illinois for using transaction fees on real estate transactions for rental housing assistance. In jail diversion, they found that Ohio has a culture of jail diversion that covers almost the entire state and Kentucky has legislation establishing a telephone triage system to screen jail inmates for mental illness and to link them to treatment. In employment and vocational success, five states, Connecticut, Maine, Missouri, New Mexico, and Vermont, had excellent ratings, while South Dakota had a deplorable 41 percent employment rate for consumers.

Oregon made a creative use of public land by reinvesting funds from the sale of a state hospital to pay for community mental health housing. California, Arizona, and Washington were cited for their multicultural outreach. Oklahoma was noted for its development of a consumer and family driven process to evaluate integrated services for co-occurring disorders. States cited for providing the best information on their Web sites were Alaska, California, Massachusetts, Michigan, Minnesota, New York, Oregon, South Carolina, Tennessee, and Texas. Model parity laws that included substance abuse were found in Connecticut, Maryland, Minnesota, Oregon, and Vermont. Maine included mental health parity in a statewide program to expand health insurance to the uninsured. Georgia was cited for its peer support program, which provides for Medicaid reimbursement of certified peer counselors.

An Ideal System

The Department of Health and Human Services through a collaborative process has developed a set of health goals and objectives for 2010. These objectives include objectives for mental health and mental disorders. The goal for mental health and mental disorders in *Healthy People 2010* is to "improve mental health and ensure access to appropriate quality mental health services" (U.S. Department of Health and Human Services 2000, 18–22), and the plan includes specific objectives:

Healthy People 2010 Objectives

Mental Health Status Improvement

1. Reduce the suicide rate to 5.0 per 100,000
   - 11.3 suicides per 100,000 population

2. Reduce the rate of suicide attempts by adolescents to 12-month average of 1 percent
   - 12-month average of 2.6 percent of adolescents in grades 9 to 12 attempted suicide in 1999
3. Reduce the proportion of homeless adults who have serious mental illness (SMI) to 19 percent
   - 25 percent of homeless adults aged 18 years and over had SMI in 1996
4. Increase the proportion of persons with serious mental illness who are employed to 51 percent
   - 43 percent of persons aged 18 years and older with SMI were employed in 1994
5. Reduce the relapse rates for persons with eating disorders including anorexia nervosa and bulimia nervosa

Treatment Expansion

6. Increase the number of persons seen in primary health care who receive mental health screening and assessment
7. Increase the proportion of children with mental health problems who receive treatment
8. Increase the proportion of juvenile justice facilities that screen new admissions for mental health problems
9. Increase the proportion of adults with mental disorders who receive treatment
10. Increase the proportion of persons with co-occurring substance abuse and mental disorders who receive treatment for both disorders
11. Increase the proportion of local governments with community-based jail diversion programs for adults with serious mental illness

State Activities

12. Increase the number of states and the District of Columbia that track consumers' satisfaction with the mental health services they receive
13. Increase the number of states, territories, and the District of Columbia with an operational mental health plan that addresses cultural competence

14. Increase the number of states, territories, and the District of Columbia with an operational mental health plan that addresses mental health crisis interventions, ongoing screening, and treatment services for elderly persons

The planning by the Department of Health and Human Services encourages planning and goal attainment by the states. It provides a basis for encouraging funding in the target areas and sets some national policy. Unfortunately, the history of long-range *Healthy People* planning shows that though some objectives end up being met, many of them are not reached. The objectives are assessed through the ten-year planning cycle. Among the findings of a 2003 assessment were that the suicide rate had fallen from 11.3 to 10.7 in 2001; the attempted suicide rate for students in grades 9 thru 12 remained 2.6; tracking consumer satisfaction with mental health services rose from thirty-six states in 1999 to thirty-nine states and the District of Columbia in 2003; and the existence of an operational mental health plan that addresses mental health crisis intervention, ongoing treatment, and mental health services for the elderly actually declined from twenty-four entities in 1997 to twenty-one states and the District of Columbia in 2002–2003 (U.S. Department of Health and Human Services 2003, 2).

Jenifer Urff (2004) found a growing consensus among public mental health professionals and consumers about an ideal system. In such a system everyone who needs treatment would have access to appropriate services. Such a system would deemphasize inpatient services for adults with serious mental illnesses and would provide comprehensive, coordinated treatment in the community. Treatment would be consumer-centered with elements that focused on providing employment, housing, peer support, and social interaction. People would receive integrated treatment for co-occurring disorders such as mental illness and substance abuse disorders, and people with mental illness who ran afoul of the criminal justice system would be diverted into the mental health system. Multidisciplinary teams would do aggressive outreach into the community, and consumers would be involved in establishing their treatment plans. Treatment plans would focus on best practices, those treatments where research has shown strong evidence of their effectiveness. Children would be identified early and receive treatment through systems of care that would integrate mental health services, education, health, and community support.

For many health professionals, the most promising new approaches involve what are called wraparound services, or integrated psychiatric networks, approaches used to care for "multiservice patients"—that is, the seriously mentally ill—to emphasize their broad needs, including housing and job training. Wisconsin, New York, and Pennsylvania use a variety of approaches involving outpatient case management and team management with round-the-clock mobile outreach teams of varying specialists and volunteers who supply therapy, medication, and advice in living skills. Since the 1970s, Dane County, Wisconsin, has pioneered an approach in which team members with diverse skills alternate in caring for persons with mental illness (Clark 1994).

Not all new approaches are expansive. In Texas, September 1, 2003, marked the beginning of a year-long statewide rollout of a new disease management approach. The approach rations ongoing outpatient care to only the most seriously mentally ill, defined as those suffering from schizophrenia, bipolar disorder, and severe clinical depression. Disease management aims to provide a continuum of care to the 89 percent of the mental health population who are listed as a priority. Along with psychiatric medication, they would receive more frequent visits, counseling, and housing and job-placement assistance. The plan is part of House Bill 2292, which overhauls social services and combines twelve agencies into five. The question remains whether this approach will help to keep people with mental illness from being hospitalized and help them to recover. Those with other mental illnesses will receive treatment only in times of crisis. The Texas executive director of the National Alliance for the Mentally Ill supported the new plan. He said that the mentally ill should learn to take personal responsibility for their recovery and work toward independent living. The business-minded system will run on less money and serve fewer clients. Its aim is to move the system toward a managed-care-provider approach based on individual consumer needs. It has been favored by the state legislature but is opposed by many mental health advocates. The executive director of the Mental Health and Mental Retardation Authority of Harris County, the largest community mental health agency in the state, which spends $52.7 million in state funds annually and operates the county psychiatric center, believes the revenues will not be sufficient to implement disease management. He believes that in his county it would take $30 million a year to fully implement disease management, but his county could only devote about one-third of its funding to that program, even with all but crisis services stopped for 530 people with diagnoses lawmakers deem less severe, including anxiety, personality disorders, and obsessive-compulsive disorders. About 4 percent of Harris County patients would lose care, while roughly 11 percent of people with mental disorders in Texas would lose care (National Alliance for the Mentally Ill 2006, 182).

The question of what an ideal system should be inevitably leads to the question of whether recovery is even a possibility. Through the history of mental illness there are cycles of optimism and pessimism about the likelihood of recovery from mental illness. Historically, recovery was the result of a particular event or treatment. It was a matter of objective improvement or apparent normalization in an individual's thought or behavior.

Recovery is now a major element in the arena of health reform. William Anthony's *Recovery from Mental Illness: The Guiding Vision of the Mental Service System in the 1990s* was published in 1993. He used the phenomenon of recovery identified by Patricia Deegan and others to formulate a plan to become a guiding vision for the provision of mental health services. He used consumers' narratives of their recovery experiences to describe processes, attitudes, values, goals, skills, feelings, and roles that lead to a satisfying and contributing life even with limitations of illness. He stated that the process of recovery can occur without professional aid, but it can be helped by the support of trusted others, and it might be correlated with a reduction in the duration and frequency of intrusive symptoms. He held that individuals needed to recover from the consequences of illness, including disability, disadvantage, and dysfunction, as much, or more, than from the illness itself. He argued that much of the disability, disadvantage, and dysfunction that people with mental illness experience is caused by the systematic and societal treatment of individuals who have a psychiatric diagnosis. He believed that two models of service provision promote recovery: the psychiatric-rehabilitation model and the community-support-system model.

The psychiatric-rehabilitation model defines and addresses the impacts of serious mental illness. The community-support-system model provides a structure for essential services in the community, including treatment, crisis intervention, and case management. Community support and psychiatric rehabilitation exist in varying degrees in communities, and Anthony emphasized the need to have a shift in attitude to a recovery model that would lead to taking intentional action to emphasize recovery-oriented mental health structures.

In the early 1990s, with declining resources for the mental health system and disillusionment by some with managed care, mental health reform was ready for a new philosophy with a values framework to balance the economic emphasis of managed care (Jacobson 2004). Rhode Island sought to add an emphasis on recovery to its managed-care program by incorporating principles such as access, innovation, quality, and cost into its system design. Ohio sponsored a conference on recovery and used it as the foundation for making changes in its system. Wisconsin used a Blue Ribbon Commission's *Final Report* in 1997 to document its implementation of a managed-care model including concepts of recovery. In her analysis of that report, Kathleen Crowley listed eight basic recovery-oriented principles that needed to be incorporated into all aspects of service delivery:

1. Recovery is possible.
2. Mental health consumers are welcomed as partners in their care.
3. A "just start anywhere" mode of consumer action is fostered.
4. A broad range of consumer-run services is promoted.
5. Meaningful work/education activities are valued and worked toward.
6. Service providers encourage and facilitate an increase in consumers' abilities to self-manage disorders.
7. Use of community resources should be encouraged.
8. Staff are empowered and encouraged to be flexible in the delivery of services. (Crowley 2000)

A Recovery Workgroup met in Wisconsin in 1998. It was composed of consumers, parents, advocates, providers, and administrators. Their charge was to operationalize the vision of recovery detailed in the Blue Ribbon Commission's report. By 1999, the Recovery Workgroup had produced two draft documents describing recommendations for operationalizing a recovery-oriented system. The Workgroup moved toward making policy recommendations at the same time that the state's managed-care model was developed. Potential barriers arose at the junction between recovery values and managed care in regard to practices and ideas such as contracting, medical necessity, and accountability (Jacobson 2004). The Recovery Workgroup was replaced by the Recovery Task Force to work on implementation. By 2003, the work of implementation was proceeding largely according to plan. A series of recovery awareness training sessions had been held in seven locations around the state to train consumers and providers.

"Anecdotally, reports of findings from the various focus groups and listening sessions held around the state during 2002—some five years after the Blue Ribbon Commission had begun talking about promoting recovery in Wisconsin—suggested that not too much had changed in many places. Consumers still lacked basic information about their diagnoses and medications, and some reported being threatened with hospitalization or loss of their children if they didn't comply with treatment plans. Providers in certain programs expressed surprise at the idea that people could recover. However, there is a growing group of consumers who are doing systems advocacy work—for example, lobbying the state legislature over recent budget cuts. Individual agencies have made changes—increasing the level of consumer involvement in organizational decision making and placing greater emphasis on client rights. It seems that change is happening but that most of it is still dependent on the work of committed individuals. It is less clear that recovery has become institutionalized, part of the warp and woof of everyday practice and policy" (Jacobson 2004, 149–150).

And it is not yet clear what recovery could mean for individuals with mental illness. How does one define recovery? And if patients are recovered, are they no longer eligible for programs such as SSI?

Recovery is also being promoted in other countries; research has been conducted in the United Kingdom and Australia, and recovery is being promoted in England. New Zealand has also been working to promote implementation of policy and practice of recovery. In the next chapter, we will examine what has happened in some other countries in regard to mental health and mental illness.

References

Aiken, L. H., S. A. Somers, and M. F. Shore (1986). Private Foundations in Health Affairs: A Case Study of the Development of a National Initiative for the Chronically Mentally Ill. *American Psychologist* 41:1290–1295.

Amers, G. (1985). Alcohol-related Movements on Drinking Policies in the American Workplace: An Historical Overview. *Journal of Drug Issues* 19:489–510.

Anderson, D. G., and M. Imle (2001). Families of Origin of Homeless and Never Homeless Women. *Western Journal of Nursing Research* 23:394–413.

Appelbaum, P. (1987). Crazy in the Streets. *Commentary* 83:34–39.

Arce, A. A., et al. (1983). A Psychiatric Profile of Street People Admitted to an Emergency Shelter. *Hospital and Community Psychiatry* 34:812–817.

Bachrach, L. L. (1987). Homeless Women: A Context for Health Planning. *Milbank Quarterly* 65:371–396.

Bachrach, L. L. (1992). What We Know about Homelessness among Mentally Ill Persons: An Analytical Review and Commentary. *Hospital and Community Psychiatry* 43:453–464.

Barker, P. R., R. W. Manderschied, G. E. Hendershot, S. S. Jack, C. Schoenborn, and I. Godstrom (1992). Serious Mental Illness and Disability in the Adult Household Population: United States, 1989. *Advance Data* 218 (September 16):1–11. Washington, DC: National Center for Health Statistics, Centers for Disease Control.

Bassuk, E. L. (1990). Who are the Homeless Families? Characteristics of Sheltered Mothers and Children. *Community Mental Health Journal* 80:1049–1052.

Bassuk, E. L., L. Rubin, and A. Lauriat (1984). Is Homelessness a Mental Health Problem? *American Journal of Psychiatry* 141:1546–1550.

Beck, A. J., and L. M. Maruschak (2001). *Mental Health Treatment in State Prisons, 2000.* Bureau of Justice Statistics Special Report, NCJ 188215. Washington, DC: U.S. Department of Justice, Office of Justice Programs.

Begun, A. M., N. R. Jacobs, and J. F. Quiram, eds. (1999). *Prisons and Jails: A Deterrent to Crime?* Wylie, TX: Information Plus.

Belcher, J. R. (1988). Rights Versus Needs of Homeless Mentally Ill Persons. *Social Work* 33:398–402.

Blum, T., J. Martin, and P. Roman (1992). A Research Note on EAP Prevalence, Components and Utilization. *Journal of Employment Assistance Research* 1:209–229.

Boardman, A. F., and D. DeMatteo (2003). Treating and Managing Sexual

Offenders and Predators. In *Correctional Mental Health Handbook,* ed. T. J. Fagan and R. K. Ax, 145–165. Thousand Oaks, CA: Sage Publications.

Bond, G. R., R. E. Drake, K. T. Mueser, and E. Latimer (2001). Assertive Community Treatment for People with Severe Mental Illness. *Disease Management and Health Outcomes* 9:141–159.

Bonta, J., M. Law, and K. Hanson (1998). The Prediction of Criminal and Violent Recidivism among Mentally Disordered Offenders: A Meta-analysis. *Psychological Bulletin* 123:123–142.

Boothroyd, R. A., N. G. Poythress, A. McGaha, and J. Petrila (2003). The Broward Mental Health Court: Process, Outcomes, and Service Utilization, *International Journal of Law and Psychiatry* 26:55–71.

Brandenberg, N., R. Friedman, and S. Silver (1990). The Epidemiology of Childhood Psychiatric Disorders: Prevalence Findings from Recent Studies. *Journal of the American Academy of Child and Adolescent Psychiatry* 35:1213–1226.

Brandes, S. D. (1970). *American Welfare Capitalism.* Chicago: University of Chicago Press.

Bumiller, E. (1999). After Attack, Giuliani Plans Crackdown on Homeless. *The New York Times,* November 20, A1, A11.

Burt, M. R., and B. E. Cohen (1989). *America's Homeless: Numbers, Characteristics, and Programs that Serve Them.* Washington, DC: Urban Institute Press. Report 89–3.

California Council of Community Mental Health Agencies (CCCMHA) (n.d.). The Development of California's Publicly Funded Mental Health System. Prepared by S. Naylor Goodwin and R. Selix. Sacramento: California Council of Community Mental Health Agencies. http://www.cccmha.org/ (accessed August 11, 2006).

California HealthCare Foundation. Majority of Mental Illness Cases Go Untreated. UCLA and Rand Corp. Study. *Archives of General Psychiatry* (January 16, 2001). http://www.californiahealthline.org/ (accessed August 11, 2006).

California HealthCare Foundation. Depression Treatment Up, with More Using Drug Therapy. California HealthCare Foundation (January 9, 2002). http://www.californiahealthline.org/ (accessed 11 August 2006).

California Healthline. Mental Health Neglected in Rural Areas. California Healthcare Foundation (November 29, 2000). http://www.californiahealthline.org/ (accessed August 11, 2006).

California Senate Rules Committee (2001) Assembly Bill Analysis-AB 88. Sacramento: California State Senate. http://info.sen.ca.gov/pub/99-00/bill/asm/ab_0051-0100/ab_88_cfa_19990909_124610_sen_floor.html (accessed September 27, 2006).

Calloway, M. O., and J. P. Morrissey (1998). Overcoming Service Barriers for Homeless Persons with Serious Psychiatric Disorders. *Psychiatric Services* 49:1568–1572.

Calsyn, R. J., R. D. Yonker, M. R. Lemming, G. A. Morse, and W. D. Klinkenberg (2005). Impact of Assertive Community Treatment and Client Characteristics on Criminal Justice Outcomes in Dual Disorder Homeless Individuals. *Criminal Behaviour and Mental Health* 15:236–248.

Camp, G. M., and C. Camp (1998). *The Corrections Yearbook, 1998*. Middletown, CT: Criminal Justice Institute.

Carter, I. (1977). Social Work in Industry: A History and a Viewpoint. *Social Thought* 3:7–31.

Caton, C. L. M., P. E. Shrout, B. Dominguez, P. F. Eagle, L. A. Opler, and F. Courmos (1995). Risk Factors for Homelessness Among Women with Schizophrenia. *American Journal of Public Health* 85:1153–1156.

Caton, C. L. M., P. E. Shrout, P. F. Eagle, L. A. Opler, A. Felix, and B. Dominguez (1994). Risk Factor for Homelessness among Schizophrenic Men: A Case-Control Study. *American Journal of Public Health* 84:265–270.

Centers for Disease Control (CDC) (1999). Scientific Data, Surveillance and Injury Statistics (May 25, 1999). http://www.cdc.gov/ncipc/osp/data.htm (accessed September 27, 2006).

Chess, S. (1969). *An Introduction to Child Psychology*. 2nd ed. New York: Grune and Stratton.

Christie, K. A., J. D. Burke Jr., D. A. Regier, et al. (1988). Epidemiologic Evidence for Early Onset of Mental Disorders and Higher Risk of Drug Abuse in Young Adults. *American Journal of Psychiatry* 145:971–975.

Clark, R.E. (1994). Family Costs Associated With Severe Mental Illness and Substance Use. *Hospital and Community Psychiatry*. 45:808–813.

Conti, R., R. G. Frank, and T. G. McGuire (2004). Insuring Mental Health Care in the Age of Managed Care. In *Mental Health Services: A Public Health Perspective*, ed. B. L. Levin, J. Petrila, and K. D. Hennessy. 2nd ed. New York: Oxford University Press.

Crowley, K. (1996). *The Power of Recovery in Healing Mental Illness*. San Francisco: Kennedy Carlisle Publishing, Health Action Network.

Dickson, W. J., and F. J. Roethlisberger (1966). *Counseling in an Organization: A Sequel to the Hawthorne Researches*. Cambridge, MA: Harvard University Press.

Diehl, S., and E. Hiland (2003). *Survey of County Jails in Tennessee: Four Years Later*. Nashville: Tennessee Department of Mental Health and Developmental Disabilities.

Diehl, S., and A. Porter (2004). *Survey of County Jails in Tennessee: One Year*

Follow Up. Nashville: Tennessee Department of Mental Health and Developmental Disabilities.

DiIulio, J. J. (1991). *No Escape: The Future of American Corrections.* New York: Basic Books.

Ditton, P. M. (1999). *Mental Health and Treatment of Inmates and Probationers.* Bureau of Justice Statistics Special Report, NCJ 174463. Washington, DC: U.S. Department of Justice: National Criminal Justice Reference Service.

Donahue, J. (1998). *The Privatization Decision: Public Ends, Private Means.* New York: Basic Books.

Drake, R. E., S. M. Essock, A. Shaner, K. B. Carey, K. Minkoff, L. Kola, D. Lynde, F. C. Osher, R. E. Clark, and L. Richards (2001). Implementing Dual Diagnosis Services for Clients with Severe Mental Illness. *Psychiatric Services* 52:469–476.

Drake, R. E., M. A. Wallach, and J. S. Hoffman (1989). Housing Instability and Homelessness Among Aftercare Patients of an Urban State Hospital. *Hospital Community Psychiatry* 40:46–51.

Drake, R. E., M. A. Wallach, G. B. Teague, D. H. Freeman, T. S. Paskus, and T. A. Clark (1991). Housing Instability and Homelessness Among Rural Schizophrenic Patients. *American Journal of Psychiatry* 148:330–336.

Empfield, M., F. Courmos, I. Meyer, et al. (1993). HIV Seroprevalence among Homeless Patients Admitted to a Psychiatric Inpatient Unit. *American Journal of Psychiatry* 150:47–52.

Essock, S. M., S. Dowden, N. T. Constantine, L. Katz, M. S. Swartz, K. G. Meador, et al. (2003). Risk Factors for HIV, Hepatitis B, and Hepatitis C among Persons with Severe Mental Illness. *Psychiatric Services* 54:836–841.

Fagan, T. J. (2003). Mental Health in Corrections: A Model for Service Delivery. In *Correctional Mental Health Handbook,* ed. T. J. Fagan and R. K. Ax, 3–19. Thousand Oaks, CA: Sage Publications.

Federal Task Force on Homelessness and Severe Mental Illness (1992). *Outcasts on Main Street.* Washington, DC: Interagency Council on the Homeless.

Feldman, D. L. (1997). The "Scarlet Letter Laws" of the 1990s: A Response to Critics. *Alabama Law Review* 60:1081–1125.

Foley, H. A., C. O. Carlton, and R. J. Howell (1996). The Relationship of Attention Deficit Hyperactivity Disorder and Conduct Disorders to Juvenile Delinquency: Legal Implications. *Bulletin of the American Academy of Psychiatry Law* 24:333–345.

Ford, C. (2005). Frequent Flyers: The High Demand User in Local Corrections. *California Journal of Health Promotion* 3, 2:61–71.

Freeman-Longo, R. E., and F. H. Knopp (1992). Sex Offender Recidivism: Issues and Outcomes. *Annals of Sex Research* 5:142–160.

Friedman, R. M., J. W. Katz-Leavy (Leavy), R. W. Manderschied, et al. (1996). Prevalence of Serious Emotional Disturbance in Children and Adolescents. In *Mental Health, United States, 1996,* ed. R. W. Manderschied and M. A. Sonnenschein. Rockville, MD: Center for Mental Health Services (CMHS).

Gelberg, L., L. S. Linn, and B. D. Leake (1988). Mental Health, Alcohol and Drug Use, and Criminal History Among Homeless Adults. *American Journal of Psychiatry* 145:191–196.

Gendreau, P. (1996). The Principles of Effective Intervention with Offenders. In *Choosing Correctional Options That Work,* ed. A. Harland, 117–130. Thousand Oaks, CA: Sage Publications.

Gendreau, P., C. Goggins, and P. Smith (1999). The Forgotten Issue in Effective Correctional Treatment: Program Implementation. *International Journal of Offender Therapy and Comparative Criminology* 43:180–187.

Grassian, S. (1983). Psychopathological Effects of Solitary Confinement. *American Journal of Psychiatry* 140:1450–1454.

Griffin, P. A., H. J. Steadman, and J. Petrila (2002). The Use of Criminal Charges and Sanctions in Mental Health Courts. *Psychiatric Services* 53:1226–1228.

Guy, E., J. J. Platt, I. Zwerling, and S. Bullock (1985). Mental Health Status of Prisoners in an Urban Jail. *Criminal Justice Behavior* 12:29–53.

Hall, G. C. N. (1995). Sexual Offender Recidivism Revisited: A Meta-Analysis of Recent Treatment Studies. *Journal of Consulting and Clinical Psychology* 63:802–809.

Haney, C., and M. Lynch (1997). Regulating Prisons of the Future: A Psychological Analysis of Supermax and Solitary Confinement. *New York Review of Law and Social Change* 23:101–195.

Harrington, S. P. M. (1997). Caging the Crazy: Supermax? Confinement under Attack. *The Humanist* 57: 14–20.

Hartwell, T. D., P. Steele, M. T. French, F. J. Potter, N. F. Rodman, and G. A. Zarkin (1996). Aiding Troubled Employees: The Prevalence, Cost, and Characteristics of Employee Assistance Programs in the United States. *American Journal of Public Health* 86:804–808.

Health and Welfare Canada (1988). *Mental Health for Canadians.* Ottawa: Minister of Supply and Services.

Herszenhorn, D. M. (1999). 1,000 in Park Denounce Giuliani's Policy of Arresting Homeless. *New York Times,* December 6, A31.

Holton, S. M. B. (2003). Managing and Treating Mentally Disordered Offenders in Jails and Prisons. In *Correctional Mental Health Handbook,* ed. T. J. Fagan and R. K. Ax, 101–122. Thousand Oaks, CA: Sage Publications.

Human Rights Watch (2003). *Ill-Equipped: U.S. Prisons and Offenders with Mental Illness.* New York: Human Rights Watch. http://www.hrw.org/reports/2003/usa1003/ (accessed September 27, 2006).

Interagency Council on the Homeless (1992). Outcasts on Main Street. HHS Pub. No. ADM 92–1904. Rockville, MD: Alcohol, Drug Abuse, and Mental Health Services Administration.

Isaac, R. J. and V. C. Armat (1990). *Madness in the Streets: How Psychiatry and the Law Abandoned the Mentally Ill.* New York: Free Press.

Jacobson, N. (2004) *In Recovery: The Making of Mental Health Policy.* Nashville, TN: Vanderbilt University Press.

Jemelka, R., E. Trupin, and J. Chiles (1989). The Mentally Ill in Prisons: A Review. *Hospital and Community Psychiatry* 40:481–491.

Jensen, G. A., M. A. Morrisey, S. Gaffney, and D. K. Liston (1997). The New Dominance of Managed Care: Insurance Trends in the 1990s. *Health Affairs* 16 (1):125–136.

Kansas v. Hendricks (117 S. Ct. 2072).

Kessler, R. C., P. A. Berglund, S. Zhao, et al. (1996). The 12-month Prevalence and Correlates of Serious Mental Illness. In *Mental Health, United States, 1996*, ed. R. W. Manderschied and M. A. Sonnenschein. DHHS Publication No. (SMA) 96–3098. Rockville, MD: CMHS.

Kessler, R. C., K. A. McGonagle, S. Zhao, et al. (1994). Lifetime and 12-month Prevalence of DSM-III-R Psychiatric Disorders in the U.S. *Archives of General Psychiatry* 51:8–19.

Kilborn, P. (1999). Gimme Shelter: Same Song, New Tune. *New York Times*, December 25, D5.

Lake, T., A. Sasser, C. Young, and B. Quinn (2002). A Snapshot of the Implementation of California's Mental Health Parity Law. Prepared for California HeathCare Foundation. Oakland, CA: Mathematica Policy Research.

Lamb, H. R., and R. W. Grant (1982). The Mentally Ill in an Urban County Jail. *Archives of General Psychiatry* 39:17–22.

Lamb, H. R., and L. E. Weinberger (1998). Persons with Severe Mental Illness in Jails and Prisons: A Review. *Psychiatric Services* 49:483–492.

Lebowitz, B. D., J. L. Pearson, and G. D. Cohen (1998). *Clinical Geriatric Psychopharmacology.* Baltimore, MD: Williams and Wilkins.

Levy, C. J. *New York Times* Investigates New York's Homes for Adults with Mental Illnesses. *New York Times* (April 28, 2002).

Levy, C. J. *New York Times* Investigates New York's Homes for Adults with Mental Illnesses. In *California Healthline* (29 April 2002). California Health Care Foundation. http://www.californiahealthline.org/ (accessed August 11, 2006).

Link, B. G., J. C. Phelan, M. Bresnahan, A. Stueve, and B. A. Pescosolido (1999). Public Conceptions of Mental Illness: Labels, Causes, Dangerousness, and Social Distance. *American Journal of Public Health* 89:1328–1333.

Lipton, D. S. (1995). *The Effectiveness of Treatment for Drug Abusers under Criminal Justice Supervision.* Washington, DC: U.S. Department of Justice, National Institute of Justice.

Lutterman, T., and R. Shaw (2000). 1999 State Mental Health Agency Profile Reports. Alexanderia, VA: National Association of State Mental Health Program Directors Research Institute.

Marshall, W. L., R. Jones, T. Ward, P. Johnston, and H. E. Barbaree (1991). Treatment Outcome with Sex Offenders. *Psychology Review* 11:465–485.

Martinson, R. (1974). What works? Questions and Answers about Prison Reform. *Public Interest* 35 (Spring): 22–35.

Mayo, E. (1923). Irrationality and Reverie. *Journal of Personnel Research* 1:477–483.

Mays, G. L., and R. Ruddell (2004). Frequent Fliers, Gang-Bangers, and Old-Timers: Understanding the Population Characteristics of Jail Populations. Paper presented at the American Society of Criminology Annual Meeting, Nashville, TN (November).

McConville, S. (1995). Local Justice: The Jail. In *The Oxford History of Prisons: The Practice of Punishment in Western Society,* ed. N. Morris and D. J. Rothman, 297–327. New York: Oxford University Press.

McRee, T., C. Dower, B. Briggance, J. Vance, D. Keane, and E. H. O'Neil (2003). The Mental Health Workforce: Who's Meeting California's Needs? San Francisco: California Workforce Initiative.

Mechanic, D. (1998). Emerging Trends in Mental Health Policy and Practice. *Health Affairs* 17:82–98.

Monahan, J., A. D. Redlich, J. Swanson, P. C. Robbins, P. S. Appelbaum, J. Petrila, H. J. Steadman, M. Swartz, B. Angell, and D. E. McNiel (2005). Use of Leverage to Improve Adherence to Psychiatric Treatment in the Community. *Psychiatric Services* 56:37–44.

Morgan, C. (1981). Developing Mental Health Services for Local Jails. *Criminal Justice Behavior* 8:259–273.

Morris, N. (1995). The Contemporary Prison. In *The Oxford History of Prisons: The Practice of Punishment in Western Society,* ed. N. Morris and D. J. Rothman, 227–259. New York: Oxford University Press.

Morse, G. A., R. J. Calsyn, G. A. Allen, B. Tempelhof, and R. A. Smith (1992). Experimental Comparison of Three Treatment Programs for

Homeless Mentally Ill People. *Hospital and Community Psychiatry* 43:1005–1010.

Morse, G. A., R. J. Calsyn, W. D. Klinkenberg, M. L. Trusty, F. Gerber, R. Smith, and E. K. Moscicki (1997). Gender Differences in Completed and Attempted Suicide. *Annals of Epidemiology* 4:152–158.

Moscicki, E. K. (2001). Epidemiology of Completed and Attempted Suicide: Toward a Framework for Prevention. *Clinical Neuroscience Research* 1:310–323.

Mueser, K. T., G. R. Bond, R. E. Drake, and S. G. Resnick (1998). Models of Community Care for Severe Mental Illness: A Review of Research on Case Management. *Schizophrenia Bulletin* 24:37–73.

Mumola, C. J. (1999). *Substance Abuse and Treatment of State and Federal Prisoners.* Bureau of Justice Statistics Special Report, NCJ 172871. Washington, DC: National Criminal Justice Reference Service.

National Advisory Mental Health Council (December 1991). *Mental Illness in America: A Series of Public Hearings.* Washington, DC: National Institute of Mental Health, National Advisory Mental Health Council.

National Advisory Mental Health Council (1993). Health Care Reform for Americans with Severe Mental Illnesses: Report of the National Advisory Mental Helath Council (1993). *American Journal of Psychiatry.* 150:1447–1465.

National Advisory Mental Health Council's Workgroup on Child and Adolescent Mental Health Intervention, Development and Deployment (2001). Blueprint for Change: Research on Child and Adolescent Mental Health. Washington, DC: National Institutes of Health.

National Alliance for the Mentally Ill (July 1999). *Families on the Brink: The Impact of Ignoring Children with Serious Mental Illness: Results of a National Survey of Parents and Other Caregivers.* Arlington, VA: NAMI.

National Alliance for the Mentally Ill of Santa Cruz County (2002). AB 1421 Passes Senate, Goes to Governor. http://www.namiscc.org/Advocacy/2002/Summer/AB1421Update.htm (accessed September 27, 2006).

National Association for the Mentally Ill (NAMI) (2006).*Grading the States 2006: A Report on America's Health Care System for Serious Mental Illness.* Arlington, VA: NAMI. http://www.nami/org/ (accessed March 2, 2006).

National Center for Health Statistics (NCHS) (1991). Centers for Disease Control and Prevention: Vital Statistics of the United States. Hyattsville, MD: NCHS.

National Center for Health Statistics (NCHS) (1992). Centers for Disease Control and Prevention (CDC). Prevalence of Chronic Circulatory Conditions. *Vital and Health Statistics* 10:94.

National Center for Missing and Exploited Children (NCMEC) Sex Offender Policy Task Force (1998). *A Model State Sex-Offender Policy.* Alexandria, VA: National Center for Missing and Exploited Children.

National Mental Health Association (2006). Can't Make the Grade: The Consequences of Cutting Mental Health Funding. http://www.nmha.org/cantmakethegrade/consequences.cfm (accessed September 3, 2006).

National Mental Health Consumers' Self-Help Clearinghouse (1999). The First National Summit of Mental Health Consumers and Survivors: Report from the Force and Coercion Plank (1999). http://www.mhselfhelp.org/pubs/view.php?publication.id=125 (accessed December 21, 2006).

Nelson, D., and S. Campbell (1990). Taylorism Versus Welfare Work in American Industry: W. J. Gault and the Bancrofts. *Business History Review* 46:1–18.

North, C. S., D. E. Pollio, B. Perron, K. M. Eyrich, and E. L. Spitznagel (2005). The Role of Organizational Characteristics in Determining Patterns of Utilization of Services for Substance Abuse, Mental Health, and Shelter by Homeless People. *Journal of Drug Issues* 22:575–591.

Office of Technology Assessment (1992). The Biology of Mental Disorders: New Developments in Neuroscience. Washington, DC: Office of Technology Assessment.

Parker, Laura (2001). Families Lobby to Force Care. *USA Today,* February 12.

Perrow, C. (1972). *Complex Organizations: A Critical Essay.* Glenview, IL: Scott, Foresman.

Pollio, D. E., E. L. Spitznagel, C. S. North, S. Thompson, and D. A. Foster (2000). Service Use Over Time and Achievement of Stable Housing in a Mentally Ill Homeless Population. *Psychiatric Services* 51:1536–1543.

Primary Mental Health Care Australian Resource Center, Department of General Practice, Flinders University and Australian Divisions of General Practice (June 2001). *Mental Health Care in Australia 2001: Report for the Commonwealth Department Health and Aged Care.* Canberra: Flinders University.

Pyle, E. (2002). New Courts Aiming to Help Mentally Ill. *The Columbus Dispatch.* Decmber 26.

Randolph, F., M. Blasinsky, J. P. Morrissey, R. A. Rosenheck, J. Cocozza, and H. H. Goldman (2002). Overview of the ACCESS Program: Access to Community Care and Effective Services and Supports. *Psychiatric Services* 53:945–948.

Regier, D.A., W. Narrow, D. S. Rae, et al. (1993). The De Facto U.S. Mental and Addictive Disorders Service System: Epidemiologic Catchment

Area Prospective 1-Year Prevalence Rates of Disorders and Services. *Archives of General Psychiatry* 50:85–94.

Ritchie, K., and D. Kildea (1995). Is Senile Dementia "Age-Related" or "Ageing Related"?—Evidence from Meta Analysis of Dementia Prevalence in the Oldest Old. *Lancet* 346:931–934.

Rochefort, D. A. (1993). *From Poorhouses to Homelessness: Policy Analysis and Mental Health Care.* Westport, CT: Auburn House.

Roethlisberger, F. J., and W. J. Dickson (1939). *Management and the Worker.* Cambridge, MA: Harvard University Press.

Rog, D., C. S. Holupka, K. L. McCombs-Thornton (1995). Implementation of the Homeless Families Program: Services, Models, and Preliminary Outcomes. *American Journal of Orthopsychiatry* 65:502–513.

Rogers, R., and M. Bagby (1992). Diversion of Mentally Disordered Offenders: A Legitimate Role for Clinicians. *Behavioral Science and the Law* 10:407–418.

Roman, P. M. (1981). From Employee Alcoholism to Employee Assistance. *Journal of Studies on Alcoholism* 42:244–272.

Roman, P. M. (1988). Growth and Transformation in Workplace Alcoholism Programming. In *Recent Developments in Alcoholism,* ed. Marc Galanter. New York: Plenum Press.

Romano, C. J. (2005). Initial Findings from the National Comorbidity Survey Replication Study. *Neuropsychiatry Reviews* 6 (6). http://www.neuropsychiatryreviews.com (accessed September 26, 2006).

Rossi, P. H. (1989). *Down and Out in America: The Origins of Homelessness.* Chicago: University of Chicago Press.

Rossi, P. H., G. A. Fisher, and G. Willis (1986). *The Condition of the Homeless of Chicago.* Chicago: National Opinion Research Council.

Rossi, P. H., and J. D. Wright (1987). The Determinants of Homelessness. *Health Affairs* 6:19–32.

Rotman, E. (1995). The Failure of Reform. In *The Oxford History of Prisons: The Practice of Punishment in Western Society,* ed. N. Morris and D. J. Rothman, 169–197. New York: Oxford University Press.

Roy, B., and R. Ruddell (2004). Diverting Mentally Ill Inmates from California Jails. *American Jails* 18:14–18.

Ruddell, R. (2006). Jail Interventions for Inmates with Mental Illnesses. *Journal of Correctional Health Care* 12:1–14.

Saez, H., E. Valencia, S. Conover, and E. Susser (1996). Tuberculosis and HIV among Mentally Ill Men in a New York City Shelter. *American Journal of Public Health* 86:1318–1319.

Schlanger, M. (2003). Inmate Litigation: Results of a National Survey. *LJN Exchange: Large Jails Network Exchange*. Annual Issue 2003, 1–12.

Shaffer, D. P., M. K. Fisher, M. K. Dulcan, et al. (1996). The NIMH Diagnostic Interview Schedule for Children, Version 2.3 (DSIC 2.3): Description, Acceptability, Prevalence Rates and Performance in the Methods for the Epidemiology of Child and Adolescent Mental Disorders Study. *Journal of the American Academy of Child and Adolescent Psychiatry* 35:856–877.

Shortage of Psychiatrists for Children Takes its Toll (2006). *Enterprise Record*, April 7, 8B.

Sonnenstuhl, W. J. (1986). *Inside an Emotional Health Program: A Field Study of Workplace Assistance for Troubled Employees*. Ithaca, NY: ILR Press.

Sonnenstuhl, W. J., and H. M. Trice (1990). *Strategies for Employee Assistance Programs: The Crucial Balance*. 2nd ed. Ithaca, NY: ILR Press.

Sosin, M. R. (2001). Service Intensity and Organizational Attributes: A Preliminary Inquiry. *Administration and Policy in Mental Health* 28:371–392.

State of Minnesota (1994). *Sex Offender Treatment Programs*. Minneapolis: State of Minnesota, Office of the Legislative Auditor, Program Evaluation Division.

Stavis, P. (2003). Mental Health's Future: Humane, Affordable. *The Times Union* (Albany, NY), January 5.

Steadman, H. J., ed. (1992). *Effectively Addressing the Mental Health Needs of Jail Detainees*. Seattle, WA: National Coalition for the Mentally Ill in the Criminal Justice System.

Steadman, H. J., S. S. Barbera, and D. A. Dennis (1994). A National Survey of Jail Mental Health Diversion Programs. *Hospital and Community Psychiatry* 45:1109–1112.

Steadman, H. J., S. Davidson, and C. Brown (2001). Law and Psychiatry: Mental Health Courts: Their Promise and Unanswered Questions. *Psychiatric Services* 52:457–458.

Steadman, H. J., K. Gounis, D. Dennis, K. Hopper, B. Roche, M. Swartz, and Clark and P. Robbins (2001). Assessing the New York City Involuntary Outpatient Commitment Pilot Program. *Psychiatric Services* 52:330–336.

Steadman, H. J., D. W. McCarty, and J. P. Morrissey (1989). *The Mentally Ill in Jail: Planning for Essential Services*. New York: Guilford Press.

Steadman, H. J., J. Monahan, E. Hartstone, S. K. Davis, and P. C. Robins (1982). Mentally Disordered Offenders: A National Survey of Patients and Facilities. *Law and Human Behavior* 6:31–38.

Steadman, H. J., S. M. Morris, and D. L. Dennis (1995). The Diversion of

Mentally Ill Persons from Jails to Community-Based Services: A Profile of Programs. *American Journal of Public Health* 85:1630–1635.

Steele, P. D. (1989). A History of Job-Based Alcoholism Programs, 1955–1973. *Journal of Drug Issues* 19:511–532.

Susser, E. S., S. P. Lin, and S. A. Conover (1991). Risk Factors for Homelessness among Patients Admitted to a State Mental Hospital. *American Journal of Psychiatry* 148:1659–1664.

Susser, E. S., E. Valencia, and S. A. Conover (1993). Prevalence of HIV Infection among Psychiatric Patients in a New York City Men's Shelter. *American Journal of Public Health* 83: 568–570.

Susser, E. S., E. Valencia, M. Miller, W. Tsai, H. Meyer-Bahlburg, and S. A. Conover (1995). Sexual Behavior of Homeless Mentally Ill Men at Risk for HIV. *American Journal of Psychiatry* 152:583–587.

Swanson, Jeffrey W., Randy Borum, Marvin S. Swartz, Virginia A. Hiday, H. Ryan Wagner, and Barbara J. Burns (2001). Can Involuntary Outpatient Commitment Reduce Arrests among Persons with Severe Mental Illness? *Criminal Justice and Behavior* 28:156–189.

Swartz, M. S., B. J. Burns, L. K. George, J. Swanson, V. A. Hiday, R. Borum, and W. H. Ryan (1997). The Ethical Challenges of Randomized Controlled Trial of Involuntary Outpatient Commitment. *Journal of Mental Health Administration* 24:35–43.

Taylor, E., O. Chadwick, E. Heptinstall, et al. (1996). Hyperactivity and Conduct Problems as Risk Factors for Adolescent Development. *Journal of the American Academy of Child and Adolescent Psychiatry* 36:1213–1226.

Teir, R., and K. Coy (1997). Approaches to Sexual Predators: Community Notification and Civil Commitment. *New England Journal on Criminal and Civil Confinement* 23:405–426.

Tempelhof, B., and L. Ahmad (1997). An Experimental Comparison of Three Types of Case Management for Homeless Mentally Ill Persons. *Psychiatric Services* 48:497–503.

Teplin, L. A. (1984). Criminalizing Mental Disorder: The Comparison Arrest Rate of the Mentally Ill. *American Psychologist* 39:794–803.

Teplin, L. A. (1990a). Detecting Disorder: The Treatment of Mental Illness among Jail Detainees. *Journal of Consulting Clinical Psychology* 58:233–236.

Teplin, L. A. (1990b). The Prevalence of Severe Mental Disorder among Male Urban Jail Detainees: Comparison with the Epidemiologic Catchment Area Program. *American Journal of Public Health* 84:290–293.

Teplin, L. A. (1994). Psychiatric and Substance Abuse Disorders among Male Urban Jail Detainees. *American Journal of Public Health* 84:290–293.

Teplin, L. A., K. M. Abram, and G. M. McClelland (1996). The Prevalence of Psychiatric Disorders among Incarcerated Women, I: Pretrial Jail Detainees. *Archives of General Psychiatry* 53:505–512.

Theriot, M. T., and S. P. Segal (2005). Involvement with the Criminal Justice System among New Clients at Outpatient Mental Health Agencies. *Psychiatric Services* 50:179–185.

Toch, H., and K. Adams (1986). Pathology and Disruptiveness among Prison Inmates. *Journal of Research in Crime and Delinquency* 23:7–21.

Torrey, E. F. (1988). *Nowhere to Go: The Tragic Odyssey of the Homeless Mentally Ill.* New York: Harper and Row.

Torrey, E. F. (1995). Jails and Prisons: America's New Mental Health Hospitals. *American Journal of Public Health* 85:1612–1620.

Torrey, E. F., E. Bargmann, and S. M. Wolfe (1985). *Washington's Grate Society: Schizophrenics in the Shelters and on the Street.* Washington, DC: Health Research Group.

Torrey, E. F., J. Stieber, and J. Ezekiel (1992). *Criminalizing the Seriously Mentally Ill: The Abuse of Jails as Mental Hospitals.* Washington, DC: Public Citizens Health Research Group and National Alliance for the Mentally Ill.

Treatment Advocacy Center (2002). Gray Matter? A Life's Work Examining Mental Disorder. Arlington, VA: Treatment Advocacy Center.

Trice, H. M., and M. Schonbrunn (1981). A History of Job-Based Alcoholism Programs, 1980–1988. *Journal of Drug Issues:* 11:171–198.

Trupin, E., and H. Richards (2003). Seattle's Mental Health Courts: Early Indicators of Effectiveness. *International Journal of Law and Psychiatry* 26:33–53.

Urff, J. (2004). Public Mental Health Systems: Structures, Goals, and Constraints. In *Mental Health Services: A Public Health Perspective.* 2nd ed., ed. B. Lubotsky Levin, J. Petrila, and K. D. Hennessy, 72–87. Oxford: Oxford University Press.

U.S. Bureau of Labor Statistics (1989). *Survey of Employer Anti-Drug Problems.* Washington, DC: Bureau of Labor Statistics, January 1989. Report 760.

U.S. Conference of Mayors (1986). *The Continued Growth of Hunger, Homelessness, and Poverty in America's Cities: 1986.* Washington, DC: United States Conference of Mayors.

U.S. Department of Health and Human Services (DHHS) (1987). *National Survey of Worksite Health Promotion Activities: A Summary.* Washington, DC: U.S. Department of Health and Human Services, Office of Disease Prevention and Health Promotion, Summer 1987. Monograph Series.

U.S. Department of Health and Human Services (DHHS) (1999). *Mental*

Health: A Report of the Surgeon General—Executive Summary. Rockville, MD: HHS, SAMHSA, CMHS, NIH, NIMH.

U. S. Department of Health and Human Services (DHHS) (2000). *Healthy People 2010: Understanding and Improving Health,* 2nd ed. Boston: Jones and Bartlett Publishers.

U. S. Department of Health and Human Services, Public Health Service (2003). *Progress Review.* Washington, DC: Health and Human Services, Public Health Service.

U.S. Department of Justice (1995). *Jails and Jail Inmates, 1993–1994.* Washington, DC: U.S. Department of Justice, Bureau of Justice Statistics. Publication NCJ-151651.

Velimesis, M. L. (1981). Sex Roles and Mental Health of Women in Prison. *Professional Psychology* 12:128–135.

Ventura, L. A., C. A. Cassel, J. E. Jacoby, and B. Huang (1998). Case Management and Atypical Antipsychotic Medications. *Psychiatric Services* 49:1330–1337.

Wallach, M. A., and J. S. Hoffman (1989). Housing Instability and Homelessness among Aftercare Patients of an Urban State Hospital. *Hospital and Community Psychiatry* 40:46–51.

Whitmer, G. E. (1980). From Hospitals to Jails: The Fate of California's Deinstitutionalized Mentally Ill. *American Journal of Orthopsychiatry* 50:65–75.

Wolff, N., R. I. Diamond, and T. W. Helminiak (1997). A New Look at an Old Issue: People with Mental Illness and the Law Enforcement System. *Journal of Mental Health Administration* 24:152–165.

Wolff, N., D. Plemmons, B. Veysey, and A. Brandli (2002). Release Planning for Inmates with Mental Illness Compared with Those Who Have Other Chronic Illnesses. *Psychiatric Services* 53:1469–1471.

Wolff, N., and J. Stuber (2002). State Mental Hospitals and Their Host Communities: The Origins of Hostile Public Reactions. *The Journal of Behavioral Health Services and Research* 29:304–317.

World Health Organization (WHO) (1946). Preamble. *Constitution.* Geneva: World Health Organization.

World Health Oranization (2001). *The World Health Report, 2001, Mental Health: New Understanding, New Hope.* Geneva: World Health Organization.

Wright, J. D., R. B. Valdez, T. Hayashi (1990). Homeless and Housed Families in Los Angles: A Study Comparing Demographic, Economic, and Family Functioning Characteristics. *Journal of Public Health* 26:425–434.

Wyatt, R. J., and E. G. DeRenzo (1986). Scienceless to Homeless. *Science* 234:1309.

3

Worldwide Perspective

World mental health is influenced by economic and political welfare. It is also influenced by geography, demography, culture, and history. Internationally, the *International Statistical Classification of Diseases and Related Health Problems (ICD)*, tenth edition, is used to define mental and behavioral disorders. The World Health Organization (WHO) since its origin has recognized the importance of mental health. The definition of health in the WHO constitution (1946) is "not merely the absence of disease or infirmity," but "a state of complete physical, mental and social well-being."

Around the world, there are people who hallucinate and have delusions, or who become depressed or anxious. The typical symptoms of mental illness seem to be recognizable in different cultural settings. Major mental health problems seem to be quite similar all over the world, including schizophrenia, depression, suicide, alcoholism, culture-specific syndromes, and mental illness among ethnic minorities and immigrants (Al-Issa 1995). Mental health care throughout the world, while endeavoring to treat many of the same disorders, sometimes differs as much in its modes of treatment as cultures differ in their values (Torrey 1986).

Mental health is crucial to the well-being of individuals and countries. However, in most countries, mental health and mental disorders are not given the same importance as physical health. "Today some 450 million people worldwide suffer from a mental or behavioral disorder, but only a small percentage of them receive even basic treatment" (World Health Organization 2001a, 3). There are also more than 800,000 people who die from suicide each year (World Health Organization 2005b). In developed countries and increasingly in developing countries, there is an increasing burden of mental disorders, with a growing number of people being untreated. Many people with mental disorders are victimized or stigmatized and discriminated against. In developing countries, most individuals with severe mental disorders have to manage within their families or alone, as there are no outside resources available to them.

The Global Burden of Mental Disorders

Four of the ten leading causes of disability worldwide are now mental disorders with 12 percent of the global burden of disease being from mental and behavioral disorders. However, mental health budgets of the majority of countries continue to constitute less than 1 percent of their total health expenditures. The World Health Organization (2001a) has predicted that there will be further increases in the number of people with mental disabilities as a result of worsening social problems, civil unrest, and aging of the population. The same report spoke of two kinds of burden created by mental problems, undefined and hidden. The undefined burden of mental problems refers to the "economic and social burden for families, communities, and countries," while the hidden burden refers to "the burden associated with stigma and violations of human rights and freedoms" (World Health Organization 2001c). Both burdens are difficult to quantify, and many cases remain unreported.

Mental illnesses have a basis in the brain and are influenced worldwide by a combination of social, biological, and psychological factors. Research throughout the world over the past twenty-five years has shown the connection between physical and mental health. One in four patients visiting a health service has at least one mental, neurological, or behavioral disorder; many of these disorders are neither diagnosed nor treated. Chronic conditions such as HIV/AIDS, cancer, diabetes, and heart disease also may affect a person's mental health, or a person's mental health may affect these chronic conditions (World Health Organization 2005b). G. M. Reed and colleagues (1994) found in a study of AIDS victims that the realistic acceptance of one's own death is associated with decreased survival time. David Spiegel and colleagues (1989) found that women with advanced breast cancer live significantly longer when they participate in supportive group therapy, and Amy Ferketich and colleagues (2000) found that depression predicts the incidence of heart disease.

The preponderance of the available research supports the hypothesis that mental and behavioral disorders are the result of both genetics and environment. The interaction of biology with psychological and social factors plays a significant role. Research also indicates that exposure to stressors during early development is associated with persistent brain hyperreactivity, which is associated with increased likelihood of depression later in life (Heim, Newport, Heit, et al. 2000). Behavior therapy for obsessive-compulsive disorder results in changes in brain function that can be observed through imaging techniques. These changes are equal to those that can be achieved by using drug therapy (Baxter, Schwartz, Bergman, et al. 1992).

Social factors such as poverty, urbanization, and technological change also play a role in the development of mental disorders, and these factors may have different results, depending on race, sex, ethnicity, and economic status (World Health Organization 2001a). Data from cross-national surveys in Brazil, Chile, India, and Zimbabwe show that common mental disorders are about twice as frequent among the poor as among the rich (Patel, Araya, Ludermir, and Todd 1999). Depression, for example, is more common among the poor than the rich. Rates of depression among rural women have been reported at twice those of the general population (Hauenstein and Boyd 1994).

"The treatment gap for most mental disorders is large, but for the poor population it is massive" (World Health Organization 2001a). Aggravating the problem, between 1950 and 2000 the proportion of urban populations in Central and South America, Africa, and Asia rose from 16 percent to 50 percent of the population (Harpham and Blue 1995, 37). Modern urbanization may have negative consequences for mental health through increased stressors and adverse life events, including reduced social support, overcrowding, poverty, pollution, and violence (Desjarlais, Eisenber, Good, and Kleinman 1995). Rural areas also create problems due to isolation, lack of transportation and communication, and limited educational and economic opportunities. Mental health services in rural areas tend to be limited or nonexistent.

Although there may be variations in the rates of specific disorders, the overall prevalence of mental disorders does not appear to be different between women and men. Antisocial personality disorders and substance use disorders are more common among men, and depressive disorders and anxiety are more common among women (Gold 1998). An estimated 121 million people worldwide suffer from depression; an estimated 5.8 percent of men and 9.5 percent of women will experience depression in any given year (World Health Organization 2001d). Depressive disorders account for 29.3 percent of disability from mental health disorders among men and nearly 41.9 percent of disability from mental health disorders among women (World Health Organization 2005a). The lifetime prevalence rate for alcohol dependence is more than twice as high in men as in women. In developed countries, approximately one in five men and one in twelve women develop alcohol dependence during their lives (World Health Organization 2005a).

There are around 24 million people worldwide who suffer from schizophrenia. Women tend to have a better outcome after treatment (World Health Organization 2001d). Schizophrenia seems to have an earlier onset and to be more disabling in men (Sartorius, Jablensky, Koretn, et al. 1986). Comorbidity is more common among women. It takes the form of a co-occurrence of anxiety, depressive, and somatoform disorders. (Somatoform disorders involve physical symptoms that are not accounted for by physical disease.) Women also report a higher number of physical and psychological symptoms than men. The prescription of psychotropic medicines is higher among women, perhaps because women are more apt to seek help. Unipolar depression is predicted to be the second leading cause of global disability burden by 2020 and it is twice as common in women, and it may be more persistent in women than men (World Health Organization 2005a). Bipolar affective disorder, on the other hand, does not show any clear difference between the sexes (Kessler et al. 1994).

> Mental and behavioral disorders are present at any point in time in about 10% of the adult population. Around 20% of all patients seen by primary health care professionals have one or more mental disorders. One in four families is likely to have at least one member with a behavioural or mental disorder. These families not only provide physical and emotional support, but also beat the negative impact of stigma and discrimination. It was estimated that, in 1990, mental and neurological disorders accounted for 10% of the total DALYs lost due to all diseases and injuries. This was 12% in 2000. By 2020,

> it is projected that the burden of these disorders will have increased to 15%. Common disorders, which usually cause severe disability, include depressive disorders, substance use disorders, schizophrenia, epilepsy, Alzheimer's disease, mental retardation, and disorders of childhood and adolescence. Factors associated with the prevalence, onset and course of mental and behavioural disorders include poverty, sex, age, conflicts and disasters, major physical diseases, and the family and social environment. (World Health Organization 2001a, 19)

In 1993, Harvard School of Public Health, in collaboration with WHO and the World Bank, assessed the global burden of disease (GBD). Estimates were made of mortality and morbidity by sex and region, and a new metric, disability-adjusted life year (DALY) was introduced to quantify the burden of disease. "One DALY is one lost year of healthy life and the burden of disease is a measurement of the gap between current health status and a healthy life where one lives into old age free of disease and disability. DALYs for a disease are the sum of the years of life lost due to premature mortality and the years lost due to disability" (Murray and Lopez 1996a, 1). Three neuropsychiatric conditions rank in the top twenty leading causes of DALYs for all ages, and six in the age group fifteen to forty-four (Murray and Lopez 1996a, 54).

Four-fifths of the world's population lives in developing regions. These regions are now in the process of dramatic change. Noncommunicable diseases such as heart disease and depression are replacing infectious diseases and malnutrition as the leading causes of disability and premature death. "By the year 2020, noncommunicable diseases are expected to account for seven out of every ten deaths in the developing regions, compared with less than half today. Injuries, both unintentional and intentional, are also growing in importance, and by 2020 could rival infectious diseases worldwide as a source of ill health" (Murray and Lopez 1996b, 1). These changes are also the result of rapid aging in the developing world. The numbers of adults relative to children is increasing as birthrates fall, causing a shift in the commonest health problems to those of adults rather than those of children.

A study found the prevalence of major psychiatric disorders in primary health care in fifteen cities showed a range for current depression from 4.0 in Shanghai, China, to 29.5 in Santiago, Chile, with a mean of 10.4. For generalized anxiety, the range was from 0.9 in Ankara, Turkey, to 22.6 in Rio de Janeiro, Brazil, with a mean of 7.8, and for alcohol dependence from 0.5 in Verona, Italy, to 7.2 in Mainz, Germany, with a mean of 2.7. The percentage of the population suffering from any mental disorder according to the Composite International Diagnostic interview ranged from 7.3 in Shanghai, China, to 35.5 in Rio de Janeiro, Brazil, with a mean of 24.0 (Goldberg and Lecrubier 1995, 327).

The burden of mental illness has been seriously underestimated by earlier approaches that focused on mortality. "Mental illnesses account for only a little over one percent of deaths, but they account for almost 11 [percent] of disease burden" (Murray and Lopez 1996b, 3). For women between the ages of fifteen and forty-four, suicide is the second leading cause of death. "In China alone, more than 180,000 women killed themselves in 1990" (Murray and Lopez 1996b, 19).

The global burden of disease (GBD) study calculated the burden of disease and injury for eight demographic regions: the established market economies (EMEs) made up largely of the Organization for Economic Cooperation and Development (OECD) countries, the former socialist economies of Europe, India, China, other Asian countries and islands, sub-Saharan Africa, Latin America and the Caribbean, and the Middle Eastern crescent. The researchers analyzed the disease burden by age groups, sex, and cause. To estimate the total burden of disability, the amount of time lived with each of various disabling injuries and diseases was calculated, in untreated and treated states. In each population, the episodes of illness were measured and weighted for their severity. Four hundred eighty-three disabling sequelae of injuries and diseases were analyzed for both sexes, all age groups, and all regions. "The GBD's findings demonstrate clearly that disability plays a central role in determining the overall health status of a population. Yet that role has until now been almost invisible to public health. The leading causes of disability are shown to be substantially different from the leading causes of death, thus casting serious doubt on the practice of judging a population's health from its mortality statistics alone" (Murray and Lopez 1996b, 21).

The study found that the burden of psychiatric disorders has been drastically underestimated. Of the ten leading causes of disability worldwide in 1990, five were psychiatric conditions: unipolar depression, alcohol use, bipolar affective disorder, schizophrenia, and obsessive-compulsive disorder. Leading causes of disability-adjusted life years (DALYs) for both sexes of all ages included, in fourth place, unipolar depressive disorders; in seventeenth place, self-inflicted injuries; and in eighteenth place, alcohol-use disorders. The leading causes of years of life lived with disability (YLDs) in all ages included the following mental disabilities: in first place, unipolar depressive disorder; in fifth place, alcohol-use disorders; in seventh place, schizophrenia; in ninth place, bipolar affective disorder; and in thirteenth place, Alzheimer's and other dementias (Murray and Lopez 1996a, 60–61). Unipolar depression was responsible for more than one in every ten years of life lived with a disability. Psychiatric and neurological conditions accounted for 28 percent of all YLDs, 1.4 percent of all deaths, and 1.1 percent of years of life lost (Murray and Lopez 1996b, 21). The burden was highest in the established market economies, but such mental disabilities were the most important contributor to YLDs in all regions except sub-Saharan Africa, where they accounted for only 16 percent of the total. In the developed regions, alcohol use is the leading cause of male disability and in the developing regions, it is the fourth leading cause of male disability. In the developed regions, alcohol use is the tenth leading cause of disability among women (Murray and Lopez 1996b, 21). "Psychiatric and neurological conditions could increase their share of the total global burden by almost half, from 10.5 per cent of the total burden to almost 15 per cent in 2020. This is a bigger proportionate increase than that for cardiovascular diseases" (Murray and Lopez 1996b, 37). In 1990, the three leading causes of disease burden were pneumonia, diarrheal disease, and perinatal conditions. By 2020, the three leading causes of disease burden are projected to be ischemic heart disease, depression, and road traffic accidents.

ICD-10 identifies two broad categories specific to childhood and adolescence: behavioral and emotional disorders and disorders of psychological development. Also many disorders more commonly found among adults can begin during childhood. The prevalence of child and adolescent disorders in children one to fifteen is 17.7 percent in Ethiopia (Tadesse, Kebede, Tegegne, and Alem 1999, 95), 21 percent in the United States (Shaffer, Fisher, Dulcan, et al. 1996, 866) and 22.5 percent in Switzerland (Steinhausen, Winkler, Metzke, et al. 1998, 265). In 2003, as part of its World Mental Health Day campaign focus on child and adolescent mental health, the World Federation for Mental Health (WFMH) drew attention to the impact of attention deficit hyperactivity disorder (ADHD) on children and their families. WFMH spearheaded an international survey of parents. ADHD is one of the most common disorders of childhood and adolescence, affecting 3–7 percent of school-aged children. The survey was conducted in nine countries (Australia, Canada, Germany, Italy, Mexico, the Netherlands, Spain, the United Kingdom, and the United States), with nearly 900 parents of children with ADHD. The study showed a wide international variability in the length of time to a diagnosis by a health professional. In the United States the average time to diagnosis is one year, while in Italy it is just over three years (World Federation for Mental Health 2004).

Suicide is among the top three causes of death in the population aged fifteen to thirty-four for both males and females. It ranks as the second cause of death for males in the European region and as the third cause of death for females. In China it ranks as the third cause of death for males in rural areas and as the first cause of death in females (World Health Organization 1994). In the Russian Federation and the Baltics, alcohol consumption has increased significantly and has been linked to an increase in rates of suicide and alcohol poisoning (Vroublevsky and Harwin 1998). These countries have also experienced a decline in male life expectancy (Leon and Shkolnikov 1998; Notzon, Komarov, Ermakov, et al. 1998).

There are a number of disorders resulting from the use of psychoactive substances including alcohol, cannabinoids such as marijuana, opioids such as opium or heroin, cocaine, hypnotics, sedatives, and volatile solvents. Alcohol is used most widely around the world and poses the most serious public health risk. There are an estimated 70 million people with alcohol-use disorders, with 78 percent untreated. The rate of alcohol use disorder for men is 2.8 percent and for women 0.5 percent. Substance-abuse disorders occur in both developed and developing countries. An estimated 5 million people worldwide inject illegal drugs, and there is a high prevalence of HIV infection among injecting drug users (World Health Organization 2001d). Traditional healers may be the only form of help available in many parts of the world. The success of folk healers and "drug bomohs" in Malaysia, based largely on the bomoh's ability to reinstill traditional values and help the person to reestablish a personal identity, demonstrates the effectiveness of traditional approaches to substance abuse (Desjarlais, Eisenber, Good, and Kleinman 1995).

For older people, Alzheimer's disease is a major threat. An estimated 37 million people worldwide have dementia, with Alzheimer's disease causing the majority of cases. About 5 percent of men and 6 percent of women over the age of sixty have Alzheimer's (World Health Organization 2001d). With the aging of the baby boom population, the number of people with Alzheimer's is expected to rise rapidly over the next twenty years. Countries throughout the world must prepare through their mental health policy and practice to deal with this and other mental health needs.

Mental Health Policy and Practice

Mental health policy is inevitably based on insufficient data, unexpected events, and unsubstantiated assumptions. Policies can only be understood within the political context in which they are developed. Various political philosophies and allegiances, values, market forces, and perceptions of need influence the compromises that power holders arrive at, and demands are made upon limited resources (Brody 1993).

At the international level, the major institutional influence on health departments and policies is the World Health Organization (WHO), the health arm of the United Nations. The Alma-Ata Declaration of 1978 called on the global community to achieve Health for All by the year 2000. Primary health care was the designated approach for achieving this goal, with mental health among its components.

The World Health Organization through its Department of Mental Health and Substance Abuse provides guidance to achieve two objectives: "closing the gap between what is needed and what is currently available to reduce the burden of mental disorders worldwide and promoting mental health" and, through its Global Action Program, "forging strategic partnerships to enhance countries' capacity to combat stigma, reduce the burden of mental disorders, and promote mental health" (World Health Organization 2005b).

More than 40 percent of countries have no mental health policy, and more than 90 percent of countries have no mental health policy for children and adolescents. Thirty percent of countries have no mental health program (World Health Organization 2001b, 3). Of 160 countries providing information on legislation, almost a fourth have no legislation on mental health. Also, nearly one-fifth of the legislation dates back over forty years (World Health Organization 2001b). In addition, health insurance often does not cover mental disorders at the same level as other illnesses.

Living conditions in psychiatric hospitals throughout the world are poor, and there are human rights violations. Human rights commissions found appalling conditions when they visited psychiatric hospitals in India (National Human Rights Commission 1999) and Central America (Levav 2000). At least one-third of the individuals in the hospitals visited did not have a psychiatric diagnosis, and case file recording was very bad. Less than half the hospitals had psychiatric social workers and clinical psychologists and less than 25 percent had psychiatric nurses. "Community care that empowers people with mental problems and the ≠development of a wide range of community services has not yet begun in many regions and countries" (World Health Organization 2001b).

In most developing countries, there is no psychiatric care for the majority of the population, and the only services available are in mental hospitals, which are used as a last resort. In many countries, laws that are more than forty years old place barriers to admission and discharge (World Health Organization 2001b). Most developing countries do not have an adequate number of trained professionals or training programs, so the community often turns to available traditional healers (Saeed, Rehman, and Mubbashar 2000). A common difference distinguishing the care of mental health patients in most developing countries from care in the United States is the extensive involvement of the family. In many developing countries, traditional healing explicitly involves family participation. For example in South Africa, a traditional healer may require the entire family to be present for a ritual. In Argentina, a *curandero* (traditional healer) uses the power of multiple family members joined in prayer to treat the patient (Susser, Collins, Schanzer, et al. 1996).

Even though, as indicated above, many countries have not yet begun to develop systems of community care, this approach is gaining in popularity around the world, with an emphasis on local and accessible care that meets a multitude of needs. The focus is on empowerment and the use of efficient treatment techniques. This kind of care occurs in some states of the United

States, some European countries, Australia, Canada, and China. Some countries in Africa, the Eastern Mediterranean, Latin America, Southeast Asia, and the Western Pacific also have some innovative services (World Health Organization 2001a).

An example of how care is changing is Saudi Arabia. Until 1983, mental health care was mainly provided by the Taif Mental Hospital, which was built for 250 patients, but by 1978 was serving 1,800 patients from all over the country.

Starting in 1983, smaller 20–120-bed hospitals are being set up all over the country, along with outpatient clinics. The next phase is to further integrate mental health with primary care (World Health Organization EMRO 2001).

The World Health Organization (2001a) defines community care as an approach that includes the following elements:

- "services which are close to home, including general hospital care for acute admissions, and long-term residential facilities in the community;
- interventions related to disabilities as well as symptoms;
- treatment and care specific to the diagnosis and needs of each individual;
- a wide range of services which address the needs of people with mental and behavioural disorders;
- services which are coordinated between mental health professionals and community agencies;
- ambulatory rather than static services, including those which can offer home treatment;
- partnership with carers and meeting their needs;
- legislation to support the above aspects of care." (p. 50)

Although psychiatric institutions with a large number of beds are no longer recommended, it is still essential to have a certain number of beds in general hospitals for acute care. There is a wide range of beds available for mental health care. The median number for the world population is 1.5 per 10,000 population, with a low of 0.33 in Southeast Asia and a high of 9.3 per 10,000 population in Europe. Nearly two-thirds of the world's population has access to fewer than one bed per 10,000, and more than half of all beds are still in psychiatric institutions that often provide custodial care rather than treatment (World Health Organization 2001b, 85).

Some countries have made major policy changes. In Italy, Law 180 in 1978 was the beginning of closing down mental hospitals. Law 180 involved exclusion of dangerousness as a criterion for commitment, restriction of compulsory hospital admission, limits on the duration of compulsory hospitalization, limits on the number of beds in psychiatric wards in civil hospitals, and sanctioning abolishment of mental hospitals. Psychiatric hospitals were closed to new admissions, and no new psychiatric hospitals were to be built. Local health authorities under regional standards implemented community services, with most services provided in the public sector, but some contracted to the private sector. The core of the service is the community mental health center with outpatient services and provision of inpatient services in small psychiatric units located inside general hospitals. The main treatment is pharmacological, and there is a quick turnover, with a mean hospitalization of less than thirteen days (Ghiradelli and Lussetti 1993).

In the mid-twentieth century, France developed the concept of the creation of geographically defined sectors, which is the main model in the organization of comprehensive psychiatric care in European countries today. The catchment areas range from 25,000 to 30,000 people. In 2005, the fifty-two member states in the WHO European Region adopted a Mental Health Action Plan for Europe (World Health Organization Europe 2005). Mental health accounts for almost 20 percent of the burden of disease in the European region. Over a lifetime, mental health problems affect one in four Europeans. Half of the people in Europe suffering from depression receive no treatment. Of the ten countries with the highest rates of suicide in the world, nine are in the European Region (World Health Organization European Region 2005).

In the Eastern Mediterranean, some developing countries have tried to develop national plans for mental health services, develop human resources, and integrate mental health with general health care. This follows the recommendations of a 1974 WHO expert committee (World Health Organization 1975).

In 1990, the World Health Organization launched an initiative for restructuring psychiatric care in the Americas, which resulted in the Declaration of Caracas. The declaration called for the development of psychiatric care linked with primary care and within the framework of the local health system.

Some countries have developed some programs of mental health care linked with primary care, including Brazil, Colombia, China, India, Iran, Pakistan, Philippines, Senegal, South Africa, and Sudan. The association of organized mental health services with primary health care provides an opportunity to avoid isolation, discrimination, and stigma. In Pakistan, general health facilities are utilized to provide mental health services. The system uses leveling of services and an efficient referral system (Mohit 2001).

The World Health Organization through Project Atlas examined the current status of mental health systems in 181 countries, covering 98.7 percent of the world's population. One-third of countries did not report a specific mental health budget, and half of the remainder allocated less than 1 percent of their public health budget to mental health, even though mental health represented 12 percent of the global burden of disease. Approximately 40 percent of the countries had no mental health policy and approximately one-third had no drug and alcohol policy (World Health Organization 2001a, 76). WHO found that this indicated a lack of "expressed commitment to address mental health problems and the absence of requirements to undertake national level planning, coordination and evaluation of mental health strategies, services and capacity" (World Health Organization 2001a, 77).

Over the past thirty years, health systems in developed countries have moved from highly centralized systems of care to decentralized systems in which responsibility for policy implementation and service provision has been transferred to local organizations. This has also occurred in some developing countries. Decentralization has occurred through reforms focused on efficiency, cost containment, and the use of contracts with private and public providers.

Experimental programs in developing countries have provided crucial models for mental health services. "Quite often, commitments from individuals, communities, and governments, coupled with minimum financial outlays, can go a long way in contributing to mental health" (Desjarlais, Eisenber, Good, and Kleinman 1995, 267).

The availability of psychotropic drugs has assisted the move to community treatment. WHO recommends a list of essential drugs for the treatment of mental disorders. Data from the Atlas

Project suggests that about 25 percent of countries do not have commonly prescribed antidepressant and antipsychotic medications available at the primary level (World Health Organization 2001a, 91). Drugs may be purchased under generic names from nonprofit organizations such as Equipment for Charitable Hospitals Overseas (ECHO) and the UNICEF Supply Division in Copenhagen. Also WHO and Management Sciences for Health issue an annual drug price guide of essential drugs that includes prices and addresses of reputable suppliers of different psychotropic drugs, at nonprofit world wholesale prices (Management Sciences for Health 2001).

Even in developed countries, utilization of available mental health services remains poor with fewer than half of those individuals needing care using available services (World Health Organization 2001a). Both inadequate services and the stigma attached to using services keeps people from using services. The U.S. Surgeon General's Report of 1999 (DHHS 1999) noted nearly half of all Americans with a severe mental illness do not seek treatment. The report noted that of the barriers to treatment, stigma was foremost in keeping people from seeking assistance. Lack of ethnically appropriate services is another barrier. Ethnicity-specific programs can increase the continued use of mental health services among ethnic minority groups. Ethnic clients who attend ethnicity-specific programs have a higher return rate and stay in treatment longer than those using mainstream services (Takeuchi, Stanley, and Yeh 1995). Barriers to effective treatment include lack of recognition of the seriousness of mental illness and lack of understanding about the benefits of services. Lack of insurance or enough money to pay for services is another barrier. Policy makers, insurance companies, health and labor policies, and the general public all make a distinction between physical and mental problems and discriminate in favor of physical problems (World Health Organization 2005b).

In a comparison of the United States and Ontario, Canada, there was an approximately 75 percent higher use of services found among persons with no morbidity or impairment in the United States compared to Ontario. Under the Ontario health system, with its universal and comprehensive insurance for mental health care, there is little evidence of excessive use of services by persons with lesser mental health problems. As there was little difference between the countries among those with higher levels of illness, the differences in prevalence of perceived need for mental health services accounted for most of the differences between the countries. Differences between U.S. citizens and Canadians may be explained by sociocultural factors, attitudes toward professional services, and adequacy of social support networks. Higher expectations and desire for medical care among Americans explain the greater use of mental health care by Americans. The difference may also be affected by the larger supply of social workers and psychologists in the United States. Even in countries with more generous insurance systems, the majority of persons with recent mental health disorders do not receive treatment. Non-insurance-related barriers remain a challenge. Also expanded coverage does not necessarily lead to a worse match between services and mental illness and impairment, as the utilization rate among persons without mental morbidity or impairment was less than 4 percent, resulting in the conclusion that there is little inappropriate use of care (Katz, Kessler, Frank, et al. 1997).

In Europe as in the United States, persons with mental illness are found in the prisons and jails. A study in England found 37 percent of men and 56 percent of women serving sentences of more than six months had a diagnosable mental disorder (McConville 1995). A study of the prevalence rates of mental illness in jails and prisons in twelve Western countries found that approximately 4 percent of their sample were diagnosed with psychotic disorders, 10 percent with major depression, and 47 percent with antisocial personality disorder (Fazel and Danesh 2002).

Financing and Human Resources

Some studies have estimated the economic costs of mental disorders. In the Netherlands, 23.2 percent of all health service costs are expenditures on mental disorders (Meerding, Bonneux, Polder, et al. 1998, 113). In the United Kingdom, inpatient expenditure only accounts for 22 percent of health spending (Patel and Knapp 1998). Most countries spend less than 3 percent of their health budgets on mental health care (Health Development Agency 2001a). Most middle and low-income countries devote less than 1 percent of their health expenditure to mental health (World Health Organization 2005b). Low treatment costs because of lack of treatment in some countries may lead to increased indirect costs resulting from large numbers of people being left untreated (Chisholm, Sekar, Kumar, et al. 2000).

The World Health Organization (2000) found that there are three factors that are essential for good financing for mental health services: protecting people from catastrophic financial risk, having the healthy subsidize the sick, and having the well-off subsidize the poor. All forms of prepayment, such as general taxation, mandatory social insurance, or voluntary private insurance, pool risks, allowing the use of services to be partly separated from direct payment for them. Chronic mental problems mean repeated services may be needed over long intervals, so what might be affordable over the short term would become prohibitive over the long term. Prepayment usually accounts for a larger share of total health spending in developed countries. While the government provides 70–80 percent of all that is spent on health in many OECD countries, in countries such as China, Cyprus, India, Lebanon, Myanmar, Nepal, Nigeria, Pakistan, and Sudan, the country only provides 20–30 percent of total financing. As those countries also have little insurance coverage, mental health is likely to suffer in comparison to other health problems, and most spending must be out-of-pocket (World Health Organization 2001a). Additional prepayment dedicated to mental and behavioral disorders can benefit mental health. Movement from prepayment to out-of-pocket spending can be detrimental, as this has been the case in several countries that used to be part of the former Soviet Union. Poor countries sometimes receive substantial assistance for health care from external donors; currently, however, donors do not give mental health high priority (World Health Organization 2001a).

In developing countries, the lack of mental health specialists is an important barrier to providing treatment. There is a wide disparity in the type and numbers of mental health workers. The median number of psychiatrists varies from 0.06 per 100,000 population in low-income countries to 9 per 100,000 in high-income countries. In almost half the world, there is less than one neurologist per million people. For psychiatric nurses, the median ranges from 0.1 per 100,000 in low-income countries to 33.5 per 100,000 in high-income countries. In developing countries, traditional healers remain the main source of assistance for at least 80 percent of rural inhabitants (World Health Organization 2001a, 95).

Family networking has proved to be beneficial in many countries, as has the development of the consumer movement. The consumer movement has influenced mental health policy in a number of countries. There is still a need in many countries to develop mental health and substance abuse policies with a wide range of stakeholders involved in the process, including consumers and their families. The World Federation for Mental Health (WFMH) was established in 1948 at the suggestion of the then interim committee for WHO and the United Nations Educational, Scientific and Cultural Organization (UNESCO). The WFMH is the oldest nongovernmental, international, and interdisciplinary organization in the mental health field. It has representation in over 100 countries. Accredited as a consultant in mental health to all of the major agencies of the United Nations, its first official recommendation was made in 1949, calling for establishment of a mental health unit within WHO. This eventually became the Division of Mental Health. The organization has advocated for rights of victims of violence, rights of mentally ill persons, and improvement of mental health care. The first annual meeting of the European Regional Council of the WFMH in Copenhagen in 1984 focused on compulsory hospitalization and treatment of persons defined as mentally ill. Results of the meeting were transmitted to the Council of Europe. The WFMH has become the leading nongovernmental organization in mental health in relation to the European Union (Brody 1993).

In the United Kingdom, the inclusion of a mental health promotion standard in the National Service Framework (NSF) for Mental Health provides for the first time a clear direction to promote mental health and to reduce the discrimination experienced by people with mental health problems. Health authorities, in conjunction with social services, were required to develop a range of strategies to meet specific performance targets by 2002:

- "Develop and agree on an evidence-based mental health promotion strategy based on local needs assessment
- Build into local strategy action to promote mental health in specific settings
- Build into local strategies action to reduce discrimination
- Ensure the written care plan for those on enhanced CPA (Care Programme Approach) shows plans to secure suitable employment or other occupational activity, adequate housing and their entitlement to welfare benefits

- Implement strategy to promote employment of people with mental health problems in health and social services." (Health Development Agency 2001b)

Unfortunately mental health in most countries still remains a low priority and much remains to be done to improve access to good community care for millions of persons with mental disabilities. Greater financial resources are needed, as well as reduction in other barriers such as insufficient mental health personnel and stigma.

References

Al-Issa, Ihsan, ed. (1995). *Handbook of Culture and Mental Illness: An International Perspective.* Madison, CT: International Universities Press.

Baxter, L. R., J. M. Schwartz, K. S. Bergman, M. P. Szuba, B. H. Guzem, J. C. Mazziotta, A. Alazraki, C. E. Selin, H. K. Ferng, and P. Munford (1992). Caudate Glucose Metabolic Rate Changes with Both Drug and Behavior Therapy for Obsessive-Compulsive Disorder. *Archives of General Psychiatry* 49:681–689.

Brody, Eugene (1993). Foreword. In *International Handbook on Mental Health Policy,* ed. Donna R. Kemp. Westport, CT: Greenwood Press.

Chisholm, D., K. Sekar, K. Kumar, K. Kishore, K. Saeed, S. James, M. Mubbashar, and R. S. Murthy (2000). Integration of Mental Health Care into Primary Care: Demonstration Cost-Outcome Study in India and Pakistan. *British Journal of Psychiatry* 176:581–588.

Desjarlais, R., L. Eisenber, B. Good, and A. Kleinman (1995). *World Mental Health: Problems and Priorities in Low-Income Countries.* New York: Oxford University Press.

Fazel, S., and J. Danesh (2002). Serious Mental Disorder in 23,000 Prisoners: A Systematic Review of 62 Surveys. *Lancet* 359:545–550.

Ferketich, A. K., J. A. Schwartbaum, D. J. Frid, and M. L. Moeschberger (2000). Depression as an Antecedent to Heart Disease among Women and Men in the NHANES I study. *Archives of Internal Medicine* 160:1261–1268.

Ghiradelli, R., and M. Lussetti (1993). Italy. In *International Handbook of Mental Health,* ed. D. Kemp. Westport, CT: Greenwood Press.

Gold, J. H. (1998). Gender Differences in Psychiatric Illness and Treatments: A Critical Review. *Journal of Nervous and Mental Diseases* 186:769–775.

Goldberg, D. P., and Y. Lecrubier (1995). Form and Frequency of Mental Disorders Across Centres. In *Mental Illness in General Health Care: An*

International Study, ed. T. B. Oxtun and N. Sartorius, 323–334. Chichester: John Wiley and Sons.

Harpham, T., and I. Blue, eds. (1995). *Urbanization and Mental Health in Developing Countries.* Aldershot, UK: Avebury.

Hauenstein, E .J., and M. R. Boyd (1994). Depressive Symptoms in Young Women of the Piedmont: Prevalence in Rural Women. *Women and Health* 21: 105–123.

Health Development Agency (2001a). International Mental Health Matters. http://www.had.nhs.uk/hdt/1101/interenational.html (accessed February 23, 2005; site now discontinued).

Health Development Agency (2001b). Mental Health Promotion: Coming in from the Cold. http://www.had.nhs.uk/hdt/0901/mental.html (accessed March 23, 2005; site now discontinued).

Heim, C., D. J. Newport, S. Heit, Y. P. Graham, M. Wilcox, R. Bonsall, A. H. Miller, and C. B. Nemeroff (2000). Pituitary-Adrenal and Autonomic Responses to Stress in Women after Sexual and Physical Abuse in Childhood. *Journal of Mental Health* 27:52–71.

Katz, Steven J., Ronald C. Kessler, Richard G. Frank, Philip Leaf, Elizabeth Lin, and Mark Edlund (1997). The Use of Outpatient Mental Health Services in the United States and Ontario: The Impact of Mental Morbidity and Perceived Need for Care. *American Journal of Public Health* 87:1136–1143.

Kessler, R. C., K. A. McGonagle, S. Zhao, C. B. Nelson, M. Huges, S. Ehleman, H. U. Wittchen, and K. S. Kendler (1994). Lifetime and 12-Month Prevalence of DSM-III-R Psychiatric Disorders in the United States: Results from the National Comorbidity Survey. *Archives of General Psychiatry* 51:9–19.

Leon, D. A., and V. M. Shkolnikov (1998). Social Stress and the Russian Mortality Crisis. *Journal of the American Medical Association* 279: 790–791.

Levav, I. (2000). Rights of Persons with Mental Illness in Central America. *Acta Psychiatrica Scandinavica* 101:83–86.

Management Sciences for Health (2001). *International Drug Price Indicator Guide.* Arlington, VA: Management Sciences for Health.

McConville, S. (1995). Local justice: The Jail. In *The Oxford History of Prisons: The Practice of Punishment in Western Society,* ed. N. Morris and D. J. Rothman, 297–327. New York: Oxford University Press.

Meerding, W. J., L. Bonneux, J. J. Polder, M. A. Koopmanschap, and P. J. van der Maas (1998). Demographic and Epidemiological Determinants of Health Care Costs in the Netherlands: Cost of Illness Study. *British Medical Journal* 317:111–115.

Mohit, Ahmad (2001). Psychiatry and Mental Health for Developing

Countries: Challenges for the 21st Century. Paper presented at the Eastern Mediterranean Region, World Health Organization, 13th Congress of Pakistan Psychiatric Society. Islamabad, Pakistan. January 25–28.

Murray, C. J. L., and A. D. Lopez, eds. (1996a). *The Global Burden of Disease: A Comprehensive Assessment of Mortality and Disability from Disease, Injuries, and Risk Factors in 1990 and Projected to 2020.* Cambridge, MA: Harvard School of Public Health for the World Health Organization and the World Bank.

Murray, C. J. L., and A. D. Lopez, eds. (1996b). *Summary: The Global Burden of Disease: A Comprehensive Assessment of Mortality and Disability from Disease, Injuries, and Risk Factors in 1990 and Projected to 2020.* Cambridge, MA: Harvard School of Public Health for the World Health Organization and the World Bank.

National Human Rights Commission (1999). *Quality Assurance in Mental Health, New Delhi:* National Human Rights Commission of India.

Notzon, F. C., Y. M. Komarov, S. P. Ermakov, C. T. Sempos, J. S. Marks, and E. V. Sempos (1998). Causes of Declining Life Expectancy in Russia. *Journal of the American Medical Association* 279:793–800.

Patel, A., and M. R. J. Knapp (1998). Costs of Mental Illness in England. *Mental Health Research Review* 5:4–10.

Patel, V., R. Araya, M. de Lima, A. Ludermir, and C. Todd (1999). Women, Poverty and Common Mental Disorders in Four Restructuring Societies. *Social Science and Medicine* 49:1461–1471.

Reed, G. M., M. E. Kemeny, S. E. Taylor, H. Y. J. Wang, and B. R. Vissher (1994). Realistic Acceptance as a Predictor of Decreased Survival Time in Gay Men with AIDS. *Health Psychology* 13:299–307.

Saeed, K., I. Rehman, and M. H. Mubbashar (2000). Prevalence of Psychiatric Morbidity among the Attendees of a Native Faithhealer at Rawalpindi. *Journal of College of Physicians and Surgeons of Pakistan* 10:7–9.

Sartorius, N., A. Jablensky, A. Koretn, C. Ernberg, M. Anker, J. E. Cooper, and R. Day (1986). Early Manifestations and First-Contact Incidence of Schizophrenia in Different Cultures. A Preliminary Report on the Initial Evaluation Phase of the WHO Collaborative Study on Determinants of Outcome of Severe Mental Disorders. *Psychological Medicine* 16:909–928.

Shaffer, D., P. Fisher, M. K. Dulcan, M. Davies, J. Piacentini, M. D. Schwab-Stone, B. B. Lahey, K. Bourdon, P. S. Jensen, H. R. Bird, C. Canino, and D. A. Regier (1996). The NIMH Diagnostic Interview Schedule for Children, Version 1.3 (DISC-2.3): Description Acceptability, Prevalence Rates, and Performance in the MECA Study. *Journal of the American Academy of Child and Adolescent Psychiatry* 35:865–867.

Spiegel, D., J. R. Bloom, H. C. Kraemer, and E. Gottheil (1989). Effect of

Psychosocial Treatment on Survival of Patients with Metastatic Breast Cancer. *Lancet* 2, 8668:888–891.

Steinhausen, H. C., C. Winkler, C. W. Metzke, M. Meier, and R. Kannenberg (1998). Prevalence of Child and Adolescent Psychiatric Disorders: The Zurich Epidemiological Study. *Acta Psychiatrica Scandinavica* 98:262–271.

Susser, Ezra, Pamela Collins, Bella Schanzer, Vijoy K. Varma, and Martin Gittelman (1996). Topics for our Times: Can We Learn from the Care of Persons with Mental Illness in Developing Countries? *American Journal of Public Health* 86:926–927.

Tadesse, B., D. Kebede, T. Tegegne, and A. Alem (1999). Childhood Behavioral Disorders in Ambo District, Western Ethiopia: I. Prevalence Estimates. *Acta Psychiatrica Scandinavica* 100 (Suppl): 92–97.

Takeuchi, David T., Sue Stanley, and May Yeh (1995). Return Rates and Outcomes from Ethnicity-Specific Mental Health Program in Los Angeles. *American Journal of Public Health* 85:638–643.

Torrey, E. Fuller (1986). *Witchdoctors and Psychiatrists.* New York: Harper and Row.

U.S. Department of Health and Human Services (DHHS) (1999). Mental Health: A Report of the Surgeon General. Rockville, MD: Department of Health and Human Services, U.S. Public Health Service.

Vroublevsky A., and J. Harwin (1998). Russia. In *Alcohol and Emerging Markets: Patterns, Problems, and Responses,* 203–223. Philadelphia: Brunner Mazel.

World Federation for Mental Health (2004). Without Boundaries—Challenges and Hopes for Living with ADHD: An International Survey. *World Federation for Mental Health Newsletter.* Third quarter.

World Health Organization (WHO) (1946). *Constitution.* Geneva: WHO.

World Health Organization (WHO) (1975). Organization of Mental Health Services in Developing Countries. Sixteenth Report of the WHO Expert Committee on Mental Health, December 1974. Geneva: World Health Organization (WHO Technical Report Series, No. 564).

World Health Organization (WHO) (1994). Cause-of-Death Statistics and Vital Rates, Civil Registration Systems and Alternative Sources of Information, 11–17. *World Health Statistics Annual 1993.* Geneva: World Health Organization.

World Health Organization (WHO) (2000). *The World Health Report 2000—Health Systems: Improving Performance.* Geneva: World Health Organization.

World Health Organization (WHO) (2001a). *The World Health Report*

2001—Mental Health: New Understanding, New Hope. Geneva: World Health Organization.

World Health Organization (WHO) (2001b). *Mental Health Resources in the World: Initial Results Of Project Atlas.* Geneva: World Health Organization (Fact Sheet No. 260).

World Health Organization (WHO) (2001c). *Mental Health Problems: The Undefined and Hidden Burden.* Geneva: World Health Organization (Fact Sheet No. 218).

World Health Organization (WHO) (2001d). *Mental and Neurological Disorders.* Geneva: World Health Organization (Fact Sheet No. 265).

World Health Organization (WHO) (2005a). *Gender and Women's Mental Health.* http://www.who.int/mental_health/prevention/genderwomen/en/print.html (accessed August 14, 2006).

World Health Organization (WHO) (2005b). *Mental Health.* http://www.who.int/mental_health/en/ (accessed August 14, 2006).

World Health Organization EMRO Regional Office for the Eastern Mediterranean (2001). *Mental Health and Substance Abuse Programme: Mental Health Programme: World Health Day.* http://www.emro.who.int/MNH/whd/CountryProfile-Saa.htm (accessed February 23, 2005).

World Health Organization Europe (2005). *Mental Health Action Plan for Europe: Facing the Challenges, Building Solutions.* WHO European Ministerial Conference on Mental Health. Helsinki, Finland. January 12–15. EUR/04/5047810/7.

World Health Organization European Region (2005). *Mental Health.* http://www.euro.who.int/mentalhealth (accessed February 23, 2005).

4

Chronology

10,000 BC In prehistoric times, there is no division between medicine, magic, and religion. Trepanning, opening the skull, is done, possibly to let out evil spirits.

400 BC The Greek physician Hippocrates treats mental illnesses as diseases to be treated in terms of disturbed physiology, rather than as displeasure of the gods.

387 BC Plato suggests that the brain is the mechanism of mental processes.

129 AD Galen, a Greek physician, writes on the theory of humors, four body fluids (blood, phlegm, choler, and black bile), whose relative proportions were believed to determine personality as well as health.

706 The first known hospital in the Islam world is built in Damascus.

1100 An asylum exclusively for people with mental diseases is established at Metz, northern France.

1284 Al-Mansuri Hospital opens in Cairo and at some time offers music therapy to its mental patients.

1290 A statute called De Praerogativa Regis gives the British king custody of the lands of the insane and of "natural fools."

1310 A German "madhouse" is established at Elbing near Danzig. Madhouses are also established in Nuremberg and Hamburg.

1349–1350 The Statute of Laborers is the first of the British poor laws which continue to affect the treatment of the mentally ill until the 1834 Poor Law Amendment.

1371 In Great Britain, a royal license is given to Robert Denton, a chaplain, to establish a hospital for the insane.

1375 The religious priory of St. Mary of Bethlem, later also known as Bedlam, in London is confiscated by the king and used for "lunatics." In 1403 a visitation report is made into the deplorable conditions at Bethlem Hospital.

1407 In Valencia, Spain, the first Spanish establishment specifically for people with mental illness is established.

1464 In Great Britain, cases occur giving custody of an "idiot" and his or her property to someone else.

1495 Syphilis, probably from the Western Hemisphere, breaks out in Italy and then throughout Europe and China. Its connection to insanity is not determined until much later.

1520 Paracelsus writes *Diseases Which Lead to a Loss of Reason*. The book makes clear that the loss of reason is not caused by spirits but by natural diseases.

1666 After the Great Fire in London, the new Bethlem (Bedlam) Hospital is built, set in gardens. Until 1770 the insane are displayed to visitors for a fee.

1670 In Britain, the earliest records of private madhouses appear. Common law is applied to madhouses: one could apply to the courts for redress of wrongful imprisonment.

1690 John Locke, in "An Essay Concerning Understanding," says madness is the inability to let reason sort out ideas by relating them correctly to experiences, setting a pattern for eighteenth-century British views that madness is a persistent inability to associate ideas correctly.

Eighteenth century Under the influence of the Enlightenment, possession by evil spirits is regarded as superstition. Psychiatry becomes an independent science.

1774 Franz Mesmer details his cure for some mental illness, originally called mesmerism and now known as hypnosis.

1789 Under the new regime established by the French Revolution, the insane are to be examined and set free or cared for in hospitals. Men are sent to Bicêtre and women to the Salpetrière. The two institutions are reserved for the insane.

1793 Philippe Pinel is appointed superintendent of Bicêtre, where he orders the unchaining of the insane. He takes charge of Salpetrière in 1795. He becomes known for providing "moral treatment."

1808 Franz Gall writes about phrenology, the belief that a person's skull shape and bumps on the head can reveal personality traits.

1824 *Elements of Phrenology* by George Combe is published.

1830 Therapeutic optimism gains ground, lasting for about thirty years. New facilities tend to be placed under the management of doctors. The belief develops that asylum treatment can provide a scientific way to cure mental illness.

1835 James Cowles Prichard's *Treatise on Insanity* becomes the main textbook on mental illness, remaining so for many years. The text focuses on moral insanity.

1838 France passes a law mandating placement of "lunatics" at public expense in every department in France. The French legal code becomes the first comprehensive legislation on mental health administration.

1841 Dorothea Dix, a mental health crusader, sees a need to treat people suffering from mental disorders and introduces the idea of psychiatric hospitals throughout the United States.

1843 The oldest surviving journal of psychiatry, *Annales Medico-psychologiques du Système Nerveux,* is founded.

1848 Phineas Gage suffers brain damage when an iron pole pierces his brain. Although his personality changes his intellect remains intact, suggesting that an area of the brain plays a role in personality.

1849 The brig *Euphemia* is converted into a prison ship in San Francisco Bay and also houses "any suspicious insane, or forlorn persons."

1850 The oldest forensic secure hospital in Europe, the Central Criminal Lunatic Asylum, later the Central Mental Hospital, is opened in Ireland. Three hospitals are established in California to treat the insane and the physically ill. The state of California assumes responsibility for the mentally disabled.

1852 California's Stockton State Hospital becomes the mental health facility for the state; it is the first hospital in the western half of the United States exclusively for the mentally disabled. It has sixty-eight patients. Therapeutic pessimism develops in the second half of the nineteenth century. It is influenced by Social Darwinism and the belief that insanity is the end product of an incurable degenerative disease inherited biologically. Reanalysis of asylum statistics leads to the belief that asylums are of little therapeutic value. They become warehouses for chronic patients.

1861 French physician Paul Broca discovers an area in the left frontal lobe that plays a key role in language development.

1869 Sir Francis Galton, influenced by Charles Darwin's *Origin of the Species,* publishes *Hereditary Genius* and argues that intellectual abilities are biological in origin.

1871 Over the next thirty years, staff shortages and overcrowding in the state institutions cause a decrease in treatment activities and more emphasis on custodial care.

1874 Carl Wernicke publishes a work on the frontal lobe that shows that damage to a specific area damages the ability to produce or understand language.

1875 In California, Napa Asylum is opened to relieve overcrowding at Stockton.

1878 G. Stanley Hall is the first American to receive a Ph.D. in psychology. He later founds the American Psychological Association.

1883 Emil Kraepelin publishes *Kompendium der Psychiatrie,* which establishes a classification of psychiatric diseases based on clusters of symptoms with underlying physical causes. He regards each mental disorder as distinct from all others. His fundamental distinction between manic depressive psychosis (bipolar) and schizophrenia holds to this day. The first laboratory of psychology in America is founded at Johns Hopkins University.

1890 New York passes the State Care Act, ordering indigent mentally ill patients out of poorhouses and into state hospitals for treatment. New York also develops the first U.S. institution for psychiatric research. The American Psychological Association (APA) is founded.

1896 The first psychological clinic is developed at the University of Pennsylvania, marking the birth of clinical psychology.

1899 Emil Kraepelin in Munich publishes *In Psychiatrie: Ein Lehrbuch Fure Studirende und Aertze,* 6th edition, classifying psychoses in two groups: dementia praecox (paranoia) and manic-depressive psychosis.

1900 Sigmund Freud publishes *The Interpretation of Dreams* as a scientific exploration of the unconscious mind. It revolutionizes psychiatric practice. He is the first to use exploration of the unconscious by free association and interpretation of dreams to treat psychiatric illness, establishing the discipline of psychoanalysis.

1905 The spirochete responsible for syphilis is identified, and the Wasserman test is established in 1906. It is first used at a British asylum in 1912, and it is calculated that one-tenth of the male population of the asylum suffers from general paralysis of the insane, that is, from syphilis-caused insanity. Sigmund Freud publishes *Three Essays on the Theory of Sexuality* describing sexual development and explaining the effects of infantile sexuality on sexual dysfunction.

1906 The *Journal of Abnormal Psychology* is founded.

1908 Clifford Beers publishes *A Mind That Found Itself,* detailing his experiences in psychiatric hospitals and leading to the founding of the mental hygiene movement in the United States.

1911 Eugene Bleuler, a Swiss psychiatrist, publishes *Dementia Praecox, oder Gruppe der Schizophrenien* in Leipzig. He coins the word *schizophrenia* in 1910 from Greek words for *split* and *mind,* as an alternative to the term *dementia praecox.* Edward Thorndike publishes the first article on animal intelligence leading to the theory of operant conditioning.

1912 Max Wertheimer publishes research which marks the

beginnings of Gestalt psychology. John Watson publishes "Psychology as a Behaviorist Views It," marking the beginnings of behavioral psychology.

1913 Carl G. Jung discontinues his Freudian views and establishes a new school of analytic psychology.

1915 The term *shell shock* is used in World War I to describe a severe neurosis originating in trauma suffered under fire.

1917 The Austrian psychiatrist Julius von Wagner-Jauregg uses malaria-induced fever to cause remission in patients with some paralysis. He becomes the first psychiatrist to win the Nobel Prize. Alfred Adler establishes the school of individual psychology and is the first psychoanalyst to challenge Freud. He coins the terms *lifestyle* and *inferiority complex* in his book, *Study of Organ Inferiority and Its Psychical Compensations.*

1918 Emil Kraepelin publishes *One Hundred Years of Psychiatry* in Germany.

1920 The Menninger Clinic (for mental health patients) is founded in Topeka, Kansas. It is named for William Menninger, who pioneered effective treatments for psychiatric casualties in World War I, and Karl Menninger, who applied psychoanalytic concepts to U.S. psychiatry.

1921 Ernst Kretschmer publishes *Korperbau und Charakter* (*Physique and Character*), arguing that body types could indicate a predisposition to mental illness.

1927 Anna Freud, daughter of Sigmund Freud, publishes her first book expanding her father's ideas for the treatment of children.

1929 In the United States, Public Law 70-672 establishes two federal "narcotics farms" and authorizes a narcotics division within the Public Health Service. The last of the large mental hospitals in the West is built in

1929 *(cont.)* Great Britain. In the United States, Public Law 71-357 redesignates the Public Health Service's Narcotics Division, making it the Division of Mental Hygiene.

1933 The Nazi Party comes to power in Germany and focuses on creating a racially pure Germany. The concern for purity leads to attempts to eliminate the mentally ill.

1935 In the United States, the Social Security Act is passed, which will eventually be a major source of funding for providing care for persons with mental disabilities. Schizophrenia is treated by inducing convulsions, first induced by the injection of camphor.

1936 Antonio Egas Moniz publishes his work on frontal lobotomies as a treatment for mental illness. Walter Freeman and James Watts develop a prefrontal lobotomy procedure, which is widely performed throughout the 1940s.

1937 Karen Horney, a German-born psychiatrist working in the United States, challenges Freud's theory of the castration complex in women and his theory that the Oedipal complex and female sexuality influence neurosis. In *The Neurotic Personality of Our Time,* she argues that neurosis is largely determined by the society in which one lives.

1938 Franz Kallman, a German-born psychiatrist working in the United States, theorizes that heredity is a relevant factor in schizophrenia. He establishes the first genetics department in a psychiatric institution in the United States. Electroshock is first used by Ugo Cerletti to produce convulsions aimed at alleviating schizophrenic and manic-depressive psychosis; later found to be more effective in the latter, it is still used. U.S. Public Law 76-19 transfers the Public Health Service from the Treasury Department to the Federal Security Agency. In Germany, Adolf Hitler decrees that patients with incurable medical conditions be killed because they are "biologically unfit." Approximately 370,000 patients with mental illness are killed. In California, outpatient care is organized within the Department of Institutions, and for the first time in California, a program is designed to help patients return to the community.

1940 Electroshock therapy is used in hospitals for the first time in the United States to treat the mentally ill.

1941 The United States begins mass production of penicillin. Its use will lead to a reduction in mental illness as a result of syphilis becoming rare.

1942 Carl Rogers publishes *Counseling and Psychotherapy,* in which he suggests that respect and a nonjudgmental approach to therapy is the foundation for the effective treatment of mental health problems.

1945 Karen Horney publishes her feminist views of psychoanalytic theory, marking the beginning of feminism in psychology.

1946 U.S. Public Law 79-487, the National Mental Health Act, authorizes the Surgeon General to improve the mental health of U.S. citizens through research into the causes, diagnosis, and treatment of psychiatric disorders. It provides generous funding for psychiatric education and research, and calls for the establishment of a National Institute of Mental Health (NIMH). It also establishes grants to states for mental health services. Anna Freud publishes *The Psychoanalytic Treatment of Children,* which introduces basic concepts in the practice of child psychoanalysis.

1949 The National Institute of Mental Health (NIMH) is established to promote research. Australian psychiatrist J. F. J. Cade introduces the use of lithium to treat psychosis. Lithium gains wide use in the 1960s to treat manic depression (bipolar).

1950 In *Childhood and Society,* Erik Erikson restates Freud's concepts of infantile sexuality and develops the concepts of adult identity and identity crisis. In California,

1950 *(cont.)* a family care program is formalized. The state pays citizens $25 per month to provide homes and foster care for state hospital patients.

1951 The French psychiatrists Jean Delay and Pierre Deniker report that chlorpromazine, later marketed under the brand name Thorazine, calms hospitalized chronic schizophrenic patients without causing significant depression. A series of successful antidepressants are introduced, and studies show that 70 percent of patients with schizophrenia improve on antipsychotic drugs.

1952 A study of psychotherapy efficacy is published by Hans Eysenck which suggests that therapy is no more effective than no treatment. This prompts a series of outcome studies which have since shown psychotherapy is an effective treatment for mental illness. *The Diagnostic and Statistical Manual of Mental Disorders* (DSM) is published by the American Psychiatric Association, marking the beginning of modern mental illness classification.

1953 The Public Health Service in the United States is assigned to the newly created Department of Health, Education, and Welfare. The U.S. psychologist B. F. Skinner publishes *Science and Human Behavior,* describing his theory of conditioning, an important concept in the development of behavior therapy.

1954 Chlorpromazine (Thorazine) becomes available as the first of the antipsychotic phenothiazines available in the United States. These drugs control symptoms but do not cure. They open the way for the deinstitutionalization movement and treatment in the community.

1955 U.S. Public Law 84-182 creates the Joint Commission on Mental Illness and Health and authorizes NIMH to study and make recommendations on mental health and mental illness in the United States. The resulting Joint Commission on Mental Illness and Health issues a report, *Action for Mental Health,* that is researched and published under the sponsorship of the thirty-six organizations making up the commission. In the United States, the number of patients in mental hospitals peaks at 560,000. A new type of therapy, behavior therapy, is developed, which holds that people with phobias can be trained to overcome them.

1957 The first effective pharmacologic treatment of depression is reported with the tricyclic antidepressant imipramine and the monoamine oxidase (MAO) inhibitor iproniazid.

1960 Martin Birnbaum, writing in the *American Bar Association Journal,* defines the right to treatment. U.S. scientists develop the benzodiazepine, chlordiazepoxide (Librium).

1961 Erving Goffman's *Asylums* is published. Goffman claims that most people in mental hospitals exhibit their psychotic symptoms and behavior as a direct result of being hospitalized. Psychiatrist Thomas Szasz publishes *The Myth of Mental Illness,* in which he argues there is no such disease as schizophrenia. *Action for Mental Health,* a ten-volume series, assesses mental health conditions and resources throughout the United States and lays out a national program to meet the needs of the mentally ill people of the United States. It is transmitted to Congress, and President Kennedy establishes a cabinet-level interagency committee to examine the recommendations and determine an appropriate federal response. In California, a ten-year plan for caring for the mentally ill leads to a decision to build no more state hospitals and to concentrate on reducing the length of treatment and enabling rapid return to the community. This decision supports development of local mental health programs.

1962 Ken Kesey's best-selling novel, *One Flew over the Cuckoo's Nest,* is based on his experiences working in the psychiatric ward of a Veterans Administration hospital. The book's premise is that patients don't really have mental illnesses, but are reacting to a rigid

1962 *(cont.)*	society. Social Security amendments allow mentally disabled individuals to receive grants. Money is available to pay for board and care in the community; 422,000 people are hospitalized for psychiatric care.
1963	President Kennedy submits to Congress the first presidential message on mental health issues. Congress quickly passes the Mental Retardation Facilities and Community Mental Health Centers Construction Act (PL 88-164), providing for grants for assistance in the construction of community mental health centers nationwide. Deinstitutionalization is mandated. This begins a new era in federal support for mental health services. American scientists develop diazepam (Valium), which comes to be widely prescribed for patients with nonpsychotic anxiety.
	Action for Mental Health recommends that the care of the mentally ill be moved from psychiatric hospitals to community mental health clinics. The California state hospital population of mentally ill peaks at 33,757 in ten state hospitals.
1965	Public Law 89-105, and amendments to Public Law 88-164, provide for grants for the staffing of community mental health centers. Congress passes the Health Insurance for the Aged Act of 1965 (Medicare) and the Grants to the States for Medical Assistance Programs Act of 1965 (Medicaid) (PL 89-97) for health programs for the poor. Both programs provide some assistance to the mentally disabled. A provision in the Social Security Amendments of 1965 provides funds and a framework for a new Joint Commission on the Mental Health of Children to recommend national action for child mental health. During the mid-1960s, NIMH launches an attack on special mental health problems. The institute establishes centers for research on schizophrenia, suicide, child and family mental health, crime and delinquency, urban problems, and minority group mental health. Later it adds aging, rape, and technical assis-

tance to victims of natural disasters. Alcohol abuse and alcoholism receive recognition as major public health problems in the mid-1960s with the establishment, as part of NIMH, of the National Center for Prevention and Control of Alcoholism. A research program on drug abuse is established within NIMH with the establishment of the Center for Studies of Narcotic and Drug Abuse.

1966 Public Law 89-793, the Narcotic Addict Rehabilitation Act of 1966, establishes a national program for long-term treatment and rehabilitation of narcotic addicts.

1967 Public Law 90-31, Mental Health Amendments of 1967, separates NIMH from the National Institutes of Health and raises it to bureau status, as well as extending construction grants to community mental health centers to cover acquisition of existing buildings. St. Elizabeth's Hospital, the federal government's only civilian psychiatric hospital, is transferred from the Department of Health, Education, and Welfare (DHEW) to NIMH. Aaron Beck publishes a psychological model of depression that suggests that thoughts play a significant role in the development and maintenance of depression.

1968 NIMH becomes a component of the newly created Health Services and Mental Health Administration (HSMHA) under the Public Health Service.

Public Law 90-574, Alcoholic and Narcotic Addict Rehabilitation Amendments of 1968, authorizes funds for the construction and staffing of new facilities for the prevention of alcoholism and the treatment and rehabilitation of alcoholics. DSM II is published by the American Psychiatric Association.

1969 Fluphenazine (Modecate), the first long-acting antipsychotic, involves one injection every few weeks, which allows for medication in the community on a

1969 *(cont.)* longer-term basis. California's Lanterman-Petris-Short Act is passed, increasing state funding to 90 percent for local community mental health programs, encouraging the trend toward community-based care and treatment. The act is hailed as a model for restoring the civil liberties of persons alleged to be mentally ill. The law establishes standards for involuntary commitment, including a first stage of a seventy-two-hour emergency hold, followed by fourteen-day and ninety-day certifications, and then conservatorships.

1970 Public Law 92-211, Community Mental Health Centers Amendments of 1970, authorizes construction and staffing of centers for three more years, with priority in poverty areas. Public Law 91-513, the Comprehensive Drug Abuse Prevention and Control Act of 1970, expands the national drug abuse program by extending services to nonnarcotic drug abusers as well as addicts and provides special project grants for drug abuse and dependence treatment programs and programs related to drug education. Public Law 91-616, the Comprehensive Alcohol Abuse and Alcoholism Prevention and Treatment and Rehabilitation Act, provides resources for special project grants for alcohol abuse and alcoholism, including a program to aid states and localities, and authorizes the establishment of a National Institute on Alcohol Abuse and Alcoholism within NIMH. The Food and Drug Administration approves lithium to treat patients with depressive illness and mania. The treatment leads to sharp drops in inpatient days and suicides and substantial savings in the economic costs associated with bipolar disorder. Dr. Julius Axelrod, an NIMH researcher, wins a Nobel Prize for research into the chemistry of nerve transmission. He has found an enzyme that terminates the action of the nerve transmitter noradrenaline in the synapse and that also serves as a critical target of many antidepressant drugs. The work of William Masters and Virginia Johnson revolutionize knowledge and attitudes about sex. They revise

Freud's theories of orgasm, report on sexual relations in geriatrics, and find counseling helps most people with sexual dysfunctions. Sex therapy becomes a psychiatric specialty. Mass deinstitutionalization begins in California. There is a lack of outpatient programs for reintegration and rehabilitation. Patients and their families have to cope.

1971 *Wyatt v. Stickney,* the first mental health institutional reform lawsuit, is filed in Alabama. The case results in court monitoring until 2003. Public Law 92-255, the Drug Abuse Office and Treatment Act of 1972, provides that a National Institute on Drug Abuse be established within NIMH.

The Social Security Amendments of 1972 establish for the first time that people who are eligible for cash benefits under the disability provision of the Social Security Act for at least twenty-four months are eligible for medical benefits.

1973 NIMH rejoins the National Institutes of Health. Then the DHEW secretary administratively establishes the Alcohol, Drug Abuse, and Mental Health Administration (ADAMHA), composed of the National Institute on Alcohol Abuse and Alcoholism, the National Institute on Drug Abuse, and NIMH. The American Psychiatric Association changes the diagnosis of homosexuality from a disease to a condition that can be considered a disease only when subjectively disturbing to the individual. Congress passes the 1973 Rehabilitation Act, which provides protections and access for people with disabilities, including mental disabilities.

1974 *R.A.J. v. Jones* lawsuit is filed in Texas. A settlement is made in 1981. The Texas mental health system is under review until 1994, when the court issues its final ruling. Public Law 93-282 authorizes the establishment of the Alcohol, Drug Abuse, and Mental Health Administration (ADAMHA).

1975 In the case *O'Connor v. Donaldson* a patient is ordered released from a mental hospital because of violation of his right to treatment. From 1970 to 1975, the population of mental illness hospitals in Britain falls from 107,977 to 87,321. The film *One Flew over the Cuckoo's Nest* dramatizes Ken Kesey's novel, showing how institutions change the personalities of people who become their inpatients. It popularizes the theories of Erving Goffman's *Asylums,* published in 1961.

1977 President Carter establishes the President's Commission on Mental Health by executive order. The commission is charged with the review of mental health needs of the nation and with making recommendations to the president as to how best to meet those needs. First Lady Rosalyn Carter is the honorary chair of the commission.

1978 Congress passes and funds a national plan for chronically mentally ill, but future-president Ronald Reagan will void funding. The four-volume report to the president from the President's Commission on Mental Health is submitted. *Rogers v. Okin* rules on the right to refuse medication. Voluntarily and involuntarily committed mental patients are assumed to be competent to manage their own affairs and cannot be forcibly medicated, except in emergency situations. When a court finds a patient to be incompetent, the judge decides if he is to be treated. Public Law 95-622, the Community Mental Health Centers Extension Act of 1978 is passed.

1979 Public Law 96-88, the Department of Education Organization Act, establishes the Department of Education and renames the Department of Health, Education, and Welfare the Department of Health and Human Services. The National Alliance for the Mentally Ill (NAMI) is founded as an organization for people with serious psychiatric illnesses and their families. They become an advocacy group for the mentally ill with legislators and the public. They provide support, education, advocacy, and research services.

1980 The Civil Rights of Institutionalized Persons Act (CRIPA) is signed into law by Jimmy Carter on May 23, 1980. The act gives the Attorney General standing to file suits against states allegedly maintaining unconstitutional conditions of confinement within state institutions such as prisons, hospitals for the mentally ill, and facilities for the mentally retarded. Public Law 96-398, the Mental Health Systems Act, reauthorizes the community mental health centers program. Based on recommendations of the President's Commission, it lays out a design to provide improved services. The Epidemiologic Catchment Area (ECA) study takes place, an unprecedented research effort involving interviews with a nationally representative sample of 20,000 Americans. Data from the ECA give an accurate picture of rates of mental and addictive disorders and services usage. The number of institutionalized mentally ill people in the United States drops from a peak of 560,000 to just over 130,000. Many people become homeless because of inadequate follow-up care and housing, and an estimated one-third of homeless people are believed to be seriously mentally ill. DSM III is published by the American Psychiatric Association.

1981 *Rennie v. Klein* rules on right to refuse medication. An involuntarily committed patient who has not been found incompetent, absent an emergency, has a qualified right to refuse psychotropic medication. Public Law 07-35, the Omnibus Reconciliation Act, repeals Public Law 96-398, the Mental Health Systems Act, and consolidates ADAMHA's treatment and rehabilitation programs into a single block grant that enables each state to administer allocated funds. The International Year for Disabled People is declared by WHO. In Great Britain, a green paper, *Care in the Community,* suggests ways of moving care and money from the National Health Service to local councils and voluntary associations. The epidemic of acquired immune deficiency syndrome (AIDS) and HIV presents mental health professionals with new challenges in treating patients' symptoms of anxiety and depression and treating symptoms of HIV brain infection. Dr. Louis

1981 *(cont.)* Sokoloff, an NIMH researcher, receives the Albert Lasker Award in Clinical Medical Research for developing a new method of measuring brain function that contributes to understanding and diagnosis of brain diseases. His technique, which measures the brain's utilization of glucose, makes possible PET scanning, the first imaging technology that permits scientists to observe the living brain.

1982 *Youngberg v. Romeo* limits the definition of right to treatment by making the standard for care "professional judgment."

1983 E. Fuller Torrey publishes *Surviving Schizophrenia: A Family Manual.* Public Law 98-24, Alcohol Abuse Amendments of 1983, consolidates the authorization for ADAMHA and the institutes into a new title V of the Public Health Service.

1984 Public Law 95-509, Alcohol Abuse, Drug Abuse, and Mental Health Amendments, authorizes funding for block grants for fiscal years 1985–1987, as well as extending the authorizations for federal activities in the areas of alcohol and drug abuse research, information dissemination, and development of new treatment methods.

1986 E. F. Torrey and S. M. Wolfe publish *Care of the Seriously Mentally Ill,* which provides the first assessment of the quality of each state's programs for the mentally ill. Advocacy groups join together to form the National Alliance for Research on Schizophrenia and Depression. It will become the largest nongovernmental, donor-supported organization for funds for brain disorder research.

1987 *Lelsz v. Kavanagh* ends federal court intervention in community mental health aftercare. G. O. Gabbard and K. Gabbard publish *Psychiatry and the Cinema,* showing the negative stereotyping of psychiatry in U.S. films. The serotonin-specific reuptake inhibitors (SSRIs) Prozac, Paxil, and Zoloft are developed by several American pharmaceutical companies to treat patients with depression. St. Elizabeth's Hospital is transferred from NIMH to the District of Columbia.

1988 John Kane, an American psychiatrist, demonstrates that clozapine is effective with schizophrenic patients who don't respond to other antipsychotics. The drug is approved by the FDA in 1989.

1989 Congress passes a resolution, subsequently signed by President George Bush, designating the 1990s as the "Decade of the Brain." The NIMH Neuroscience Center and the NIMH Neuropsychiatric Research Hospital are dedicated in Washington, D.C.

Early 1990s Scientists find the genetic variations that cause Fragile X syndrome, early-onset Alzheimer's disease, and Huntington's disease.

1990 The Human Genome Project is established. Brain imaging is used to learn more about the development of mental illness. Congress passes the Americans with Disabilities Act, which provides protections for people with disabilities, including those with mental disabilities.

1991 Public Law 99-550, the Public Health Service Act, contains the requirement for state comprehensive mental health services plans. In California, realignment reorganizes authority and control over resources for counties, with a focus on creating a single system of care.

1992 Public Law 102-321, the ADAMHA Reorganization Act, abolishes ADAMHA, creates the Substance Abuse and Mental Health Services Administration (SAMHSA), and transfers NIMH research activities to the National Institutes of Health. New offices are created for research on prevention, special populations, rural mental health, and AIDS. A survey of American jails reports that 7.2 percent of inmates are seriously mentally ill, meaning that 100,000 seriously mentally ill people are in jails and prisons. Over a quarter of

1992 *(cont.)* them are held without charges, often awaiting a bed in a psychiatric hospital.

1993 NIMH establishes the Human Brain Project to develop, through imaging and computer and network technologies, a comprehensive neuroscience database accessible via an international computer network. A review of neuroimaging studies indicates that three brain regions are involved in schizophrenia: the frontal, the temporolimbic, and the basal ganglia.

1994 The FDA approves risperidone, another drug that is effective in treating refractory schizophrenic patients and is the first new first-line antipsychotic drug in almost twenty years. The Veterans Health Programs Extension Act of 1994 adds the treatment of sexual trauma and places no limitation on the time to seek services. DSM IV is published by the American Psychiatric Association.

1995 The Social Security Administration becomes an independent agency.

1996 NIMH, with NAMHC (National Advisory Mental Health Council), initiates systematic reviews of several research areas, including the genetics of mental disorders; epidemiology and services for child and adolescent populations; prevention research; and clinical treatment and services research. Childhood mental disorders are given priority by NIMH. NIMH establishes implementation of *Human Subjects Protection in Clinical Research* to expand its efforts to safeguard and improve protections for human subjects who participate in clinical mental health research. Congress passes the Mental Health Parity Act of 1996, requiring that annual and lifetime dollar limits on mental health care not be stricter than for other medical care.

1997 Researchers identify genetic links to bipolar disorder. NIMH realigns and establishes three research divisions: Basic and Clinical Neuroscience Research;

Services and Intervention Research; and Mental Disorders, Behavioral Research, and AIDS.

1998 Psychology advances to the technological age with the emergence of e-therapy, which allows mental health treatment to be provided over the internet.

1999 The Ticket to Work and Work Incentives Improvement Act of 1999 modernizes employment services for peopled with disabilities, making it possible for Americans with disabilities to work without fear of losing their Medicaid or Medicare coverage. The first White House Conference on Mental Health is held in Washington, D.C., and brings together national leaders, mental health, scientific, and clinical personnel, patients, and consumers to discuss needs and opportunities. NIMH convenes its fourth rural mental health research conference, "Mental Health at the Frontier: Alaska," in Anchorage. The aim is to solicit assistance in the development of a research agenda focusing on mental health issues for people who live in rural or frontier areas, with a focus on Alaska Natives. NIMH hosts "Dialogue: Texas," the first in a series of mental health forums to solicit input from the public on the direction of future research and NIMH and to highlight current research. The Surgeon General releases *The Surgeon General's Call to Action To Prevent Suicide* and the first Surgeon General's Report on Mental Health. California joins other states in passing mental health parity legislation requiring insurance companies to provide insurance coverage for mental health comparable to that for other medical conditions.

2000 Human genome sequencing is published. NIMH continues to strengthen its efforts to include the public in its priority-setting and strategic-planning processes, including members of the public on its scientific review committees reviewing grant applications. NIMH launches a five-year communications initiative, the Constituency Outreach and Education Program, to disseminate science-based mental health information

2000 *(cont.)* to the public and health professionals and increase access to effective treatment. Public Law 106-310, the Children's Health Act of 2000, Title 1 Autism, expands and coordinates activities with respect to research on autism, including the establishment of not less than five centers of excellence to conduct basic and clinical research into autism.

2001 The report, *The Surgeon General's National Action Agenda for Children's Mental Health,* is released. The report indicates the nation is facing a public crisis in the mental health of children and adolescents. The National Action Agenda outlines goals and strategies to improve services for children and adolescents with mental and emotional disorders. In Pittsburgh, NIMH convenes more than 150 scientists with expertise in the study of mood disorders to help develop a Research Strategic Plan for Mood Disorders. A public forum held in conjunction with the meeting focuses on the frequent co-occurrence of depression with general medical illnesses. NIMH launches several long-term, large-scale, multisite community-based clinical studies to determine the effectiveness of treatment for bipolar disorder; depression in adolescents; antipsychotic medications in the treatment of schizophrenia; management of psychotic symptoms and behavioral problems associated with Alzheimer's disease; and subsequent treatment alternatives to relieve depression. The World Health Organization releases a landmark publication aimed at raising public and professional awareness of the real burden of mental disorders, *The World Health Report 2001: Mental Health: New Understanding, New Hope.* The New York Academy of Medicine is the first to assess the mental health, substance use, and respiratory health impact of the September 11 terrorist attacks.

2002 NIMH releases a report, *Mental Health and Mass Violence: Evidence-Based Early Psychological Intervention for Victims/Survivors of Mass Violence.* The report indicates that early psychological intervention guided by qualified mental health caregivers can reduce the harmful psychological and emotional effects of exposure to mass violence in survivors. New Mexico becomes the first state to pass legislation allowing licensed psychologists to prescribe psychotropic medication.

2003 NIMH launches the Real Men, Real Depression campaign to raise awareness about depression in men and create an understanding of the signs, symptoms, and treatment available.

2004 In California, Proposition 63 is passed, which provides funding for mental health programs by levying a one percent income tax on all California taxpayers with an annual taxable income of more than $1 million.

2005 The theme of World Mental Health Day 2005 was Mental Health for Life. SAMHSA estimated that 500,000 people had significant health problems after Hurricanes Katrina and Rita.

2006 New York and federal health officials announced they would conduct another survey of people who lived through the events of September 11, 2001. Seventy-one thousand people who enrolled in the World Trade Center Health registry will be be surveyed regarding the psychological and environmental stressors to which they were exposed. An earlier survey in 2003 to 2004 reported thousands of people still reported significant mental health and respiratory impacts.

5

Biographies

Karl Abraham (1877–1925)

A leader in psychoanalysis, he was one of Freud's earliest and most important collaborators. He was a member of the Vienna Psychoanalytic Circle and a founder of the Berlin Institute. He made important contributions to the theory of psychosexual development and influenced many of the psychoanalysts whom he analyzed at the Berlin Institute.

Nathan W. Ackerman (1908–1971)

A pioneer of family therapy, he studied medicine at Columbia University and later psychiatry. In 1955 he helped found the American Academy of Psychoanalysis. His appreciation of social and cultural determinants led to his treatment of the family as a group. In 1960 he founded the Family Institute, New York, and, with cofounder Don Jackson, the journal *Family Process.* He specialized in the practice and teaching of family therapy and wrote numerous articles and books.

Alfred Adler (1870–1937)

Founder of individual psychology, Adler joined the Vienna Psychoanalytic Circle founded by Freud. He broke with Freud in 1911 to develop his own approach to psychoanalysis. He founded the Society of Individual Psychology and several child guidance clinics. His approach became known as Adlerian psychology. He was the author of several books, including *The Practice of Individual Psychiatry,* published in 1924.

Franz Gabriel Alexander (1891–1964)

Alexander studied medicine at the University of Budapest and was appointed to the Hygiene Institute of Budapest. During World War I he became a military physician. After the war, he studied psychoanalysis at the Berlin Psychoanalytic Institute and became an assistant there. In 1929 he became the first professor of psychoanalysis at the University of Chicago's Department of Medicine, and in 1931 he founded the Chicago Psychoanalytic Society and served as its first director. He published important studies of the personality and the application of psychoanalytic theory to criminality. In 1946 he published his most important work, *Psychoanalytic Therapy.* He emphasized the importance of the corrective emotional experience with therapy, and he proposed the possibility of briefer psychoanalysis. His work was met with protest, and he completed his career as a professor at the University of Southern California.

Gregory Bateson (1904–1980)

An anthropologist, communication theorist, and systems thinker, he married Margaret Mead in 1936. Bateson's work centered on communication and schizophrenia. From 1954 to 1959 he was director of a research project on schizophrenic communication. In 1956, with Jay Haley, Don Jackson, and John Weakland, he published the seminal work *Toward a Theory of Schizophrenia.* He also worked on communication and family psychotherapy, for which he was widely recognized in the family therapy field.

Clifford W. Beers (1876–1943)

A graduate of Yale University and businessman who had attempted suicide and been hospitalized for manic-depressive psychosis from 1900 to 1903, he vowed to expose hospital conditions and work for their reform. In 1908 the result was a book, *A Mind That Found Itself.* He also was a founder of the mental health movement and was instrumental in establishing the National Committee for Mental Hygiene in 1909. In 1913 he founded the first outpatient mental health clinic in the United States.

Eric Berne (1910–1970)

Founder of transactional analysis, he received psychoanalytic training in New York, where he started working with groups. He moved to Carmel, California, where he worked on ego states. He published the first paper on transactional analysis in 1958. An International Association of Transactional Analysis was formed in 1964. Berne focused on the nature of transactions and classified the variety of games that individuals play in communication. He then developed his ideas on scripts, which map out adult actions based upon childhood occurrences. Berne wrote numerous books, including *Games People Play,* published in 1964.

Eugene B. Brody (1921–)

Dr. Eugene B. Brody completed his medical degree at Yale University, and in 1959 he was appointed as the second chair of the Department of Psychiatry at the University of Maryland School of Medicine. In 1976, Dr. Brody resigned as chair, but continued as a professor at the university. From 1985–1986 he was a Sackler Scholar. He also served as editor in chief of the *Journal of Nervous and Mental Disease.* He served a term as president of the World Federation of Mental Health, and then was for a number of years secretary general of the World Federation, where he engaged in numerous activities to promote mental health. He retired as secretary general in 1999, but continues as senior consultant to that organization. Currently Dr. Brody is Professor and Chairman Emeritus of Psychiatry, University of Maryland; Visiting Professor of Psychiatry and Senior Advisor for Refugee Trauma, Harvard Medical School; and Senior Advisor at the Johns Hopkins School of Public Health.

George W. Bush (1946–)

During George Bush's governorship in Texas, the Texas Medication Algorithm Project (TMAP) was established as a model medication treatment plan using evidence-based practice to determine medication needs of persons with mental illness. The Texas project started in 1995 as an alliance of individuals from the pharmaceutical industry, the University of Texas, and the mental health and corrections systems of Texas. A Robert Wood Johnson grant and several drug companies funded the project. During his 2000 presidential campaign, Bush promoted his support for the project and the fact that legislation he had proposed had expanded Medicaid coverage of psychotropic drugs. When Bush became president, he established the New Freedom Commission on Mental Health in April 2002 to conduct a comprehensive study of the mental health system. The commission issued its recommendations in July 2003. The commission found that mental disorders often go undiagnosed. They recommended comprehensive mental health screening including for preschool children. The commission felt schools were in a position to screen the 52 million students in school and the 6 million adults working in the schools. They also recommended linking screening with treatment and encouraged the use of science-based treatments with designated medications for specific conditions. Bush instructed more than twenty-five federal agencies to develop an implementation plan based on those recommendations.

Rosalyn Carter (1927–)

The wife of President Jimmy Carter, she was actively involved in promoting community mental health centers during the time her husband was governor of Georgia. One of the first acts of President Carter was to create a Commission on Mental Health with his wife as honorary chairwoman. She continues to work on mental health issues through the Carter Center's Mental Health Program.

Jean Martin Charcot (1825–1893)

Director of the Salpetrière Hospital in Paris, he established the most famous neurological clinic of the nineteenth century, and he was recognized as outstanding as both a teacher and a clinician because of his ability to relate symptoms to the nervous system. His major research was conducted on neuroses. Freud came to Paris to study hypnosis under Charcot.

Bill Clinton (1946–)

In his 1999 State of the Union address, President Clinton said that the country needed to step up its efforts to treat and prevent mental illness and that "no American should ever be afraid . . . to address this disease." The Clinton administration worked to improve mental health treatment, enhance prevention, support research, and reduce stigma and discrimination. It advocated for and signed into law the 1996 Mental Health Parity Act (MHPA). In December 1997, the administration issued regulations to implement this law to end discrimination in health insurance on the basis of mental illness. As of January 1998, the law began requiring health plans to provide the same annual and lifetime spending caps for mental health benefits as they do for medical and surgical benefits. The president also worked for a 1997 Balanced Budget Act that included $24 billion to provide health care coverage to millions of uninsured children with a strong mental health benefit as part of the program. During the Clinton administration, the Surgeon General prepared the first *Surgeon General's Report on Mental Health,* which was released in 1999. The document reviewed the most current science and recommended approaches for promoting mental health, preventing mental illness, and providing state-of-the-art clinical interventions across the life cycle. Also in 1999, President Clinton signed an executive order ensuring that individuals with psychiatric disabilities were given the same work opportunities as persons with other disabilities.

Albert Deutsch (1905–1961)

As a researcher of historical documents for the New York Department of Welfare, he became interested in the problems of the mentally ill. In 1937 he published a history, *The Mentally Ill in America,* which established him as a leading writer on mental illness. As a journalist, in 1949 he wrote *The Shame of the States,* an indictment of the appalling conditions in state mental hospitals.

Albert Ellis (1913–)

In 1955, Albert Ellis developed rational emotive behavior therapy (REBT), which teaches individuals to replace their self-defeating thoughts and beliefs with more effective life-enhancing ideas and behaviors. The theory has elements in common with cognitive and behavior therapy. It employs imagery, confrontation, and structured exercises to enable the client to experience himself differently. It aims to both eliminate symptoms and restructure the personality. Ellis founded the Albert Ellis Institute in 1955, and he has published numerous books.

Erik Homburger Erikson (1902–1979)

Born in Denmark, he moved at an early age to Germany. As an art teacher in Vienna, he helped teach the children of patients being analyzed by Sigmund and Anna Freud, and he entered analysis with Anna Freud. He completed his analytic training at the Vienna Psychoanalytic Institute in 1933. He left Europe during the rise of Nazi Germany. At Harvard University he worked with Gregory Bateson and Margaret Mead, as well as with psychologists Kurt Lewin and Henry Murray. In 1939 he moved to California, where he continued his analytic work with children and researched children's play. He integrated ideas from psychoanalysis, anthropology, and psychiatry. He developed a theory of human epigenesis as involving a continuing exploration of identity throughout the life cycle, and he researched the essential influence of social and cultural factors on the formation of the ego. He also applied insights of psychoanalysis to the lives of great historical figures, writing psychological biographies of Gandhi and Luther. He founded the Erikson Institute for Early Childhood Education in Chicago.

Anna Freud (1895–1982)

The youngest of Freud's children, she was a pioneer in child analysis and child psychotherapy. She published her seminal work, *The Ego and the Mechanisms of Defense,* in 1937. In 1939, she and her father escaped to England from the Nazis, and she remained there, living and working in London. She founded the Hampstead Child Therapy Course in 1947 and the Hampstead Clinic in 1952.

Sigmund Freud (1856–1939)

The father of American and European psychiatry, he vastly extended the scope of behavior considered appropriate for doctors to treat, explaining mental illness in terms of the same psychodynamic processes as normal behavior. The founder of psychoanalysis, he was born in Moravia and moved as a child to Vienna. He received his M.D. from the University of Vienna in 1881 and had a period of residency at the Viennese General Hospital. He then went to Paris to study hypnosis under Jean Charcot. In 1902, he was appointed professor at the University of Vienna and formed the first Psychoanalytic Society. He and his early followers, Jung, Adler, Wilhelm Stekel, and Sandor Ferenczi, explored and developed the major areas of psychoanalytic study, including the study of infantile sexuality, the unconscious, and dreams. Many of Freud's discoveries were published in the form of case studies. He wrote numerous books, and the standard edition of his works was published by the Hogarth Press, London. Near the end of his life, he suffered thirty surgeries for cancer of the jaw, and he fled Vienna for exile in London.

Erich Fromm (1900–1980)

Fromm received his doctorate from the University of Heidelberg in 1922. He took his psychoanalytic training at the Berlin Psychoanalytic Institute. He emigrated to the United States in 1933 and taught at the New School for Social Research in New York. He became a neo-Freudian, with a strong interest in sociocultural influences and the effect of social and political conditions on personality development. Some of his best known books are *The Art of Loving* (1956) and *Escape from Freedom* (1941).

William Glasser (1925–)

William Glasser was trained in psychiatry, and during his psychiatry residency at the Veterans Administration Hospital in Los Angeles from 1954 to 1957, he began his break with traditional psychiatry. Along with G. L. Harrington, he developed an approach to therapy that he named reality therapy. He conducted a private practice from 1957 to 1986 using his concept of reality therapy. It emphasized the need to help patients face reality and fulfill their basic human needs. It had links with both cognitive therapy and learning theory. He founded the Institute for Reality Therapy in 1967, which was renamed the Institute for Control Theory, Reality Therapy, and Quality Management in 1994, and the William Glasser Institute in 1996. By the 1970s, he called his body of work control theory and that changed to choice theory in 1996. Glasser does not believe in the concept of mental illness unless there is something organically wrong with the brain that can be determined by a pathologist. Very early he concluded that genetically people are social creatures that need each other, and that the cause of almost all psychological symptoms is the inability to get along with the important people in a person's life. Over the years he has developed ten axioms of choice theory and has written numerous books including *Reality Therapy: A New Approach to Psychiatry, Positive Addiction, Schools Without Failure, The Identity Society, Mental Health or Mental Illness?*, and *Stations of the Mind.* At his institute, he has trained numerous psychotherapists in his approach.

Erving Goffman (1922–1982)

Erving Goffman was the author of *Asylums*, published in 1961 and based on fieldwork done in 1955–1956 at St. Elizabeth's Hospital in Washington, D.C. His book was strongly antipsychiatric, arguing that mental hospitals were only one of a number of "total institutions," including army barracks, jails, and boarding schools. Goffman accused the institution of causing the deviant behavior that it is meant to cure. Goffman ignored the neuroleptic drugs because they reinforced the medical model that he had discarded. Although he had astute perceptions of patterns of interaction and the functioning of the institution, he had no sense of where the institution was going.

Mike Gorman (1948–1989)

A journalist turned mental health lobbyist, he published a book, *Every Other Bed,* in which he revealed that half of all hospital beds in the United States were occupied by the mentally ill and a quarter of the beds were held by people diagnosed as schizophrenic. With public health advocates and philanthropists Mary Lasker and Florence Mahoney, he moved from the local level to the national scene as a mental health reform lobbyist. He also furthered the establishment of patient advocacy groups.

Carl Gustav Jung (1875–1961)

An early disciple of Freud, Jung was the originator of analytic psychology. Jung conducted early experiments with word association, and in 1907 he became a member of the Vienna Psychoanalytic Circle. In 1911 he became the first president of the International Psychoanalytic Society. Although Freud originally considered him his heir apparent, by 1914 Jung's views had diverged widely from Freud, and Jung broke his association with Freud. His approach became known as Jungian psychology. Jung was a prolific writer, and his *Collected Works* consist of twenty volumes. He concentrated on understanding the individual as a whole person, in his spiritual and collective aspects as well as in terms of his biological and individual needs.

John F. Kennedy (1917–1963)

As president, John F. Kennedy had a personal interest in mental health, as he had a sister with mental retardation and mental illness. Upon taking office in 1961, he directed that top-level discussions begin in the Department of Health, Education, and

Welfare to develop a response to the Joint Commission's 1961 report, *Action for Mental Health.*

The administration's approach went beyond the recommendations of the Joint Commission and focused on the establishment of a wholly new, independent community system composed of autonomous centers offering a full range of services. Under the new policy, all federal resources would be directed toward community mental health centers. In 1962, President Kennedy established the President's Interagency Task Force on Mental Health, with the charge of drafting legislation to propose a new program to Congress. This work led eventually to the Community Mental Health Centers Act.

R. D. Laing (1927–1989)

Born in Glasgow in 1927, he studied medicine at Glasgow University. In 1953 he set up an experimental "therapeutic community" for twelve of the most disturbed chronic schizophrenic patients on his ward in a Scottish hospital. Laing's first book, *The Divided Self,* published in 1960, marked a departure in that, while many psychiatrists had sought to explain why the individual became ill by looking at the family, he treated the behavior itself as intelligible, a "rational strategy" in the face of a terrifying family environment. Sales of the book took off after Laing allied himself with the New Left and became a cult figure. In 1967 he published *The Politics of Experience,* in which he wrote, "There is no such 'condition' as schizophrenia but the label is a social fact and the social fact *a political event.*" Laing was a major figure in the antipsychiatry movement.

Abraham Harold Maslow (1908–1970)

One of the most influential leaders of humanistic psychology, he was born in New York City. He studied psychology at the University of Wisconsin, receiving his degree in 1934. World War II had a profound effect on Maslow, and he began to develop his own theory of human motivation, emphasizing the normality of human growth and studying the hierarchies of human needs in terms of problem solving, perception, and cognition. This led him to his major work on self-actualization, peak experiences, and the development of a humanistic approach to psychotherapy. He took a major part in the founding of the Esalen Institute at Big Sur, California, the first center for the human potential movement. Among his important books are *Motivation and Personality* in 1954 and *Toward a Psychology of Being* in 1962.

David Mechanic (1936–)

David Mechanic is the author of more than 400 publications, including 20 books, among which are *Future Issues in Health Care: Social Policy and the Rationing of Medical Services, From Advocacy to Allocation: The Evolving American Health Care System,* and *Mental Health and Social Policy.* He is the Rene Dubois University Professor of Behavioral Sciences and director of the Institute for Health, Health Care Policy and Aging Research at Rutgers. He serves as the director of the Robert Wood Johnson Foundation Investigator Awards Program in Health Policy Research. He has received the Distinguished Investigator Award from the Association for Health Services Research and numerous other awards from the American Public Health Association and the American Sociological Association.

Fritz Perls (1891–1970)

Fritz Perls was born in Berlin, studied at the Berlin and Vienna institutes of psychoanalysis, and was analyzed by Wilhelm Reich, who was trained as a medical doctor, psychiatrist, and Freudian psychoanalyst. He later immigrated from Austria to the United States. In 1934, to escape Nazism, Perls went to South Africa where he founded the South African Institute of Psychoanalysis. By 1947, when he published his first book, it was clear that a new system of psychotherapy was emerging from his thought. With the rise of apartheid in South Africa, Perls moved to the United States. With his wife Laura, he founded the New York Institute for Gestalt Therapy. The term *Gestalt therapy* was first used as the title of a book published by Perls, Ralph Hefferline, and Paul Goodman in 1951. Gestalt therapy is a noninterpretative psychotherapy, which emphasizes awareness and personal responsibility and adopts a holistic approach, giving equal emphasis to mind and body. It began to achieve prominence toward the end of the

1960s and is now one of the leading psychotherapies in the United States. In 1960 Perls moved to California and conducted much of his work at the Esalen Institute.

Otto Rank (1894–1939)

Otto Rank was born in Vienna and was a member of Freud's inner circle. He particularly applied Freudian concepts to the interpretation of art, mythology, and literature. His most original work, *The Trauma of Birth,* was published in 1923. He broke with Freud and left Vienna in 1924. He settled in the United States and developed the concept of will therapy, which emphasized the individual's will as serving as a positive guide and integrating the self. He founded the Pennsylvania School of Social Work, which propagated his work. He had a great influence on humanistic psychology and on experiential and existential therapy.

Ronald Reagan (1911–2004)

President Reagan's approach to the seriously mentally ill was predictable. As governor of California, he presided over a massive deinstitutionalization of patients from state mental hospitals and a shift of fiscal responsibility from the state to the counties. Deinstitutionalization failed in California, as it did in other states. But there was no acknowledgment of that failure when Reagan reached the White House. Some of his domestic policies, such as sharply reducing incentives for builders to put up low-income housing and trying to reduce federal Supplemental Security Income (SSI) payments to the disabled, exacerbated the problems of the seriously mentally ill.

B. F. Skinner (1904–1990)

Perhaps the most well-known of the behaviorists, B. F. Skinner studied behavior modification, which consists of various methods and techniques derived from the principles of learning theory. He developed what came to be known as Skinnerian principles based on operant conditioning theory. Behavior modification techniques include techniques used in inpatient settings, such as contingency management and the development of a token economy. Skinner even suggested the use of behavior modification as a way of running whole societies. Biofeedback techniques are also used to assess problems, and the cognitive and affective levels of the patient are usually not addressed directly, as the patient-therapist relationship is not considered to be an important treatment variable. Behavior modification has been criticized on ethical grounds for its manipulative approach.

Thomas Szasz (1910–)

Born in Hungary and educated in the United States, Szasz published *The Myth of Mental Illness* in 1961. Szasz offered a variety of "proofs" that mental illness did not exist. In his book *Schizophrenia,* Szasz treated mental illness as an invention of psychiatrists. Szasz is a major figure in the antipsychiatry movement.

E. Fuller Torrey (1937–)

E. F. Torrey has authored or edited more than twenty books including *Nowhere to Go: The Tragic Odyssey of the Homeless Mentally Ill, Surviving Schizophrenia,* and *The Roots of Treason: Ezra Pound and the Secret of St. Elizabeth's.* He is instrumental in the development and annual publication of a rating system of state mental health programs for the care of the seriously mentally ill. The National Alliance for the Mentally Ill and the Public Citizen Health Research Group publish these reports. Each state is ranked according to a set of criteria and given a letter grade from A to F. He is an advocate of involuntary outpatient treatment, and a specialist on clinical and research work on schizophrenia. He has researched viruses as a possible cause of some mental disorders.

6

Facts and Statistics

Much of the data available in the mental health area is government data. Some of the data, such as suicide rates, can be found by looking at general health care, but most of the data can be found directly under mental health. Much of the federal data is now relatively easy to find because it is on the Internet. Also hard copies of data and reports can be found at libraries that are designated as government repository libraries. These libraries used to be the major source of such data, but now are not so important, as much of the data can be found directly on the Web. Some of the data can be found through looking at Web sites of nonprofit organizations.

Almost all federal agencies have their own Web sites, which display various types of government data. There is a general Web site for finding the federal government at http://www.firstgov.gov/. There is also a specialized Web site for government statistics at http://www.fedstats.gov, which has a section on general health statistics. In the first part of this chapter, information is provided on some of the major data sets and types of data available at the national level, including which agencies maintain which data. Information on specific states is also found through some federal government Web sites. Some states may have information available through state or local government Web sites, but this chapter will not attempt to cover those sites. A good overview of finding and using mental health statistics is found in the University of Chicago's John Carter Library's "Health Statistics Research Guides—Research Guide to Mental Health Statistics," located at http://www.lib.uchicago.edu/e/su/med/healthstat/mental.html.

Federal Agencies and Mental Health Data Collections

The most important source for mental health statistics is the Center for Mental Health Services (CMHS). A wide array of mental health statistics may be found at the CMHS. The Center is in the Substance Abuse and Mental Health Services Division of the Department of Health and Human Services.

Center for Mental Health Services (CMHS)

CMHS is the only program in the country that focuses on the development of data standards on mental health services. They seek to provide uniform and comparable statistics. This approach is very important, as you can examine data across states and compare one state with another. It is the model in the mental health care statistics field. The center provides assistance to governments on improving the quality of mental health information systems in order to enhance the quality of mental health programs and services. The center can be found at http://mentalhealth.samhsa.gov/cmhs/MentalHealthStatistics/default.asp.

CMHS has collected data since 1840, which gives it one of the longest continuous data collections in public health. This means that data can be compared over time. For example, change could be looked at from the 1970s, when the consumer movement began, to the present. CMHS developed the National Reporting Program for Mental Health Statistics, through which it collects data. It then analyzes and reports national statistical information on mental health services and the people served. "Program activities include:

- Biennial enumeration surveys of all specialty mental health organizations in the U.S.
- Periodic targeted client sample surveys of persons served by specialty mental health organizations.
- Special surveys of mental health services in nontraditional settings, such as prisons and jails, juvenile justice settings, and mental health self-help activities.
- The National Conference on Mental Health Statistics, an annual event that provides a state-of-the-art examination

of the technologies and methodologies of data collection for representatives of State, community, academic, consumer, and family groups.

- An examination of indirect indicators of the risk of mental illness through a Health Demographic Profile System and the direct assessment of health status through collaborative studies." (Center for Mental Health Statistics 2005)

The National Reporting Program is a joint effort of national and local organizations. Data are collected on mental health organizations, their clients, finances, and staff, making the program the authoritative data source on mental health organizations and allowing for comparison of different mental health organizations. The results of the National Reporting Program activities are available in a biennial publication on major mental health policy and statistical issues, *Mental Health, United States.*

CMHS also develops standards that make national mental health data collection uniform and comparable throughout the United States. Through the Mental Health Statistics Improvement Program, CMHS provides "guidance and technical assistance to decision makers at all levels of government on the design, structure, content, and use of mental health information systems to improve the quality of mental health programs and service delivery" (CMHS 2005). The program celebrated its thirtieth anniversary in May 2006. It is a collaborative effort of state mental health agencies and other key groups. The collaborative nature of the program makes it a model in the health care statistics field.

Under the Mental Health Statistics Improvement Program, you can obtain a consumer-focused mental health report card on managed care developed by CMHS. The report card presents performance indicators recommended for assessing mental health services to adults in the areas of prevention, access, quality, and outcomes.

CMHS has completed a State Performance Indicator Pilot Project to improve the reporting capability of state mental health data systems and to establish a basis for comparing information about performance indicators across states. "Under the project, 16 State Mental Health Authority grantees tested 32 performance indicators and measures for State mental health programs, with a focus on outcomes. Examples of indicators included, level of functioning and symptoms, use of assertive community treatment services, use of supported employment services, and use of atypical medications" (CMHS 2005). The lessons learned from this project will be used to help with future state planning and decision making.

State Mental Health Statistics

"State statistics are provided for each state for the following areas:

> *Adults with Serious Mental Illness—2002.* The estimated number of adults within the State who were suffering from mental illness in the year 2002.
>
> *Children and Adolescents with Serious Emotional Disturbance—2002.* The estimated number of children and adolescents within the State who were suffering from serious emotional disturbances in the year 2002.
>
> *State Mental Health Block Grant Funding—2002.* The dollar amount received by the State from the Federal Government through the mental health block grant program.
>
> *State Mental Health Agency.* Mental Health Actual Dollar and Per Capita Expenditures, 2001.
>
> *State and County Psychiatric Hospital, Inpatient Beds—1998.* The number of beds in publicly funded psychiatric hospitals in the State in 1998.
>
> *State and County Psychiatric Hospitals, Inpatient Census—2000.* The number of inpatients in publicly funded psychiatric hospitals in the State at the end of 2000." (CMHS 2005)

The Center for Mental Health Services (CMHS) is the source for the mental health statistics in Tables 6.1 through 6.6. Table 6.1 provides information for each of the states and the District of Columbia on the number of persons with serious mental illness in each state in 2002. The numbers are for adults only, eighteen and older, and are the number of residents in the state and the number of residents in the state with a serious mental illness. The statistics use 5.4 percent as the percentage of persons with a serious mental illness but also include a lower estimate at 3.7 percent and a higher estimate at 7.1 percent. The table allows comparisons across states. The discrepancy in the estimates highlights one of

TABLE 6.1
Number of Persons with Serious Mental Illness, Aged 18 and Older, by State, 2002
Does not Include Persons Who Are Homeless or Institutionalized

State	Resident population (2002)	Resident population with SMI (5.4%)	Lower limit of estimate (3.7%)	Upper limit of estimate (7.1%)
Alabama	3,379,400	182,488	125,038	239,937
Alaska	451,358	24,373	16,700	32,046
Arizona	3,979,597	214,898	147,245	282,551
Arkansas	2,032,557	109,758	72,205	144,312
California	25,663,642	1,385,837	949,555	1,822,119
Colorado	3,355,424	181,193	124,151	238,235
Connecticut	2,587,650	139,733	95,743	183,723
Delaware	617,687	33,355	22,854	43,856
District of Columbia	458,770	24,774	16,974	32,573
Florida	12,830,878	692,867	474,742	910,992
Georgia	6,291,833	339,759	232,798	446,720
Hawaii	949,384	51,267	35,137	67,406
Idaho	970,692	52,417	35,916	68,919
Illinois	9,346,097	504,689	345,806	663,573
Indiana	4,564,211	246,467	168,876	324,059
Iowa	2,238,715	120,891	82,832	158,949
Kansas	2,019,365	109,046	74,717	143,375
Kentucky	3,161,303	170,710	116,968	224,453
Louisiana	3,296,972	178,036	121,988	234,085
Maine	1,015,406	54,832	37,570	72,094
Maryland	4,078,212	220,223	150,894	289,553
Massachusetts	4,964,461	268,081	183,685	352,477
Michigan	7,480,182	403,930	276,767	531,093
Minnesota	3,767,595	203,450	139,401	267,499
Mississippi	2,111,035	113,996	78,108	149,883
Missouri	4,275,118	230,856	158,179	303,533
Montana	693,133	37,429	25,646	49,212
Nebraska	1,289,787	69,648	47,722	91,575
Nevada	1,600,901	86,449	59,233	113,664
New Hampshire	966,685	52,201	35,767	68,635
New Jersey	6,462,909	348,997	239,128	458,867
New Mexico	1,354,553	73,146	50,118	96,173
New York	14,544,281	785,391	538,138	1,032,644
North Carolina	6,251,306	337,571	231,298	443,843
North Dakota	487,298	26,314	18,030	34,598

continues

TABLE 6.1 *(continued)*
Number of Persons with Serious Mental Illness, Aged 18 and Older, by State, 2002
Does not Include Persons Who Are Homeless or Institutionalized

State	Resident population (2002)	Resident population with SMI (5.4%)	Lower limit of estimate (3.7%)	Upper limit of estimate (7.1%)
Ohio	8,541,340	461,232	316,030	606,435
Oklahoma	2,620,154	141,488	96,946	186,031
Oregon	2,666,408	143,986	98,657	189,315
Pennsylvania	9,471,639	511,469	350,451	672,486
Puerto Rico	2,714,765	146,597	100,447	192,747
Rhode Island	830,477	44,846	30,728	58,964
South Carolina	3,128,020	168,913	115,737	222,089
South Dakota	565,438	30,534	20,921	40,146
Tennessee	4,392,628	237,202	162,557	311,877
Texas	15,677,577	846,589	580,070	1,113,108
Utah	1,603,244	86,575	59,320	113,830
Vermont	476,930	25,754	17,646	33,862
Virginia	5,514,134	297,763	204,023	391,504
Washington	4,555,636	246,004	168,559	323,450
West Virginia	1,412,702	76,286	52,270	100,302
Wisconsin	4,103,132	221,569	151,816	291,322
Wyoming	376,359	20,323	13,925	26,721

Source: Center for Mental Health Services, http://mentalhealth.samhsa.gov/cmhs/MentalHealthStatistics/default.asp.

the issues in determining the number of people with mental illness in the United States. There are no exact figures. Various research studies have provided information on people with mental illness, but the figures vary. Thus the Center for Mental Health Services chooses to use three different figures to project the number of people with mental illness in the states—a mid-range percentage and a high and low percentage. Accurate figures are not obtainable because many people with serious mental health problems are not being treated, so the numbers cannot be obtained from treatment records, and community studies rely on self-reporting. The figures do not include the homeless mentally ill, for whom there is a wide array of estimates, nor do they include those who are institutionalized. As the number of persons with mental illness is reflected as a percentage of the resident population of the state, California, with the largest resident population, has the largest population of persons with mental illness, with a low estimate of 949,555, a mid-range estimate of 1,385,837, and a high estimate of 1,822,119.

Table 6.2 estimates for the year 2002 the number of children and adolescents with serious emotional disturbance by state. The

TABLE 6.2
Estimated Number of Children and Adolescents with a Serious Emotional Disturbance, by State, 2002

			Level of functioning		Level of functioning	
State	Number of youth aged 9 to 17	Percent in poverty	50 Lower limit	Upper limit	60 Lower limit	Upper limit
Alabama	571,345	17.5%	28,567	39,994	51,421	62,848
Alaska	103,091	10.2%	5,155	7,216	9,278	11,340
Arizona	732,232	17.4%	36,614	51,260	65,905	80,551
Arkansas	348,271	27.7%	17,414	24,379	31,344	38,310
California	4,801,547	18.4%	240,077	336,108	432,139	528,170
Colorado	583,369	13.3%	29,168	40,836	52,503	64,171
Connecticut	454,669	10.0%	22,733	31,827	40,920	50,014
Delaware	97,915	11.2%	4,896	6,854	8,812	10,771
District of Columbia	53,355	34.8%	2,668	3,735	4,802	5,869
Florida	2,016,048	16.0%	100,802	141,123	181,444	221,765
Georgia	1,132,144	15.2%	56,607	79,250	101,893	124,536
Hawaii	148,947	14.0%	7,447	10,426	13,405	16,384
Idaho	191,326	12.9%	9,566	13,393	17,219	21,046
Illinois	1,647,318	15.4%	82,366	115,312	148,259	181,205
Indiana	817,153	9.9%	40,858	57,201	73,544	89,887
Iowa	368,883	9.6%	18,444	25,822	33,199	40,577
Kansas	359,970	10.6%	17,999	25,198	32,397	39,597
Kentucky	478,472	18.9%	23,924	33,493	43,062	52,632
Louisiana	613,369	22.6%	30,668	42,936	55,203	67,471
Maine	157,180	17.2%	7,859	11,003	14,146	17,290

continues

TABLE 6.2 *(continued)*
Estimated Number of Children and Adolescents with a Serious Emotional Disturbance, by State, 2002

State	Number of youth aged 9 to 17	Percent in poverty	Level of functioning 50 Lower limit	Upper limit	Level of functioning 60 Lower limit	Upper limit
Maryland	718,356	7.7%	35,918	50,285	64,652	79,019
Massachusetts	759,502	13.6%	37,975	53,165	68,355	83,545
Michigan	1,351,713	14.0%	67,586	94,620	121,654	148,688
Minnesota	665,876	7.4%	33,294	46,611	59,929	73,246
Mississippi	389,840	21.4%	19,492	27,289	35,086	42,882
Missouri	733,407	13.7%	36,670	51,338	66,007	80,675
Montana	119,451	16.9%	5,973	8,362	10,751	13,140
Nebraska	228,421	10.9%	11,421	15,989	20,558	25,126
Nevada	282,005	10.9%	14,100	19,740	25,380	31,021
New Hampshire	168,889	6.1%	8,444	11,822	15,200	18,578
New Jersey	1,088,400	8.3%	54,420	76,188	97,956	119,724
New Mexico	262,991	23.2%	13,150	18,409	23,669	28,929
New York	2,380,138	19.1%	119,007	166,610	214,212	261,815
North Carolina	1,037,951	18.7%	51,898	72,657	93,416	114,175
North Dakota	79,701	15.4%	3,985	5,579	7,173	8,767
Ohio	1,496,274	9.8%	74,814	104,739	134,665	164,590
Oklahoma	449,748	18.0%	22,487	31,482	40,477	49,472
Oregon	444,534	12.9%	22,227	31,117	40,008	48,899
Pennsylvania	1,541,625	13.1%	77,081	107,914	138,746	169,579
Rhode Island	126,917	13.2%	6,346	8,884	11,423	13,961
South Carolina	508,281	16.9%	25,414	35,580	45,745	55,911
South Dakota	104,139	10.9%	5,207	7,290	9,373	11,455
Tennessee	718,154	17.4%	35,908	50,271	64,634	78,997
Texas	3,051,443	20.8%	152,572	213,601	274,630	335,659
Utah	340,566	11.6%	17,028	23,840	30,651	37,462
Vermont	79,002	11.4%	3,950	5,530	7,110	8,690
Virginia	911,922	13.0%	45,596	63,835	82,073	100,311
Washington	792,689	12.0%	39,634	55,488	71,342	87,196
West Virginia	208,613	23.8%	10,431	14,603	18,775	22,947
Wisconsin	717,987	11.7%	35,899	50,259	64,619	78,979
Wyoming	67,280	9.9%	3,364	4,710	6,055	7,401

Source: Center for Mental Health Services, http://mentalhealth.samhsa.gov/cmhs/MentalHealthStatistics/default.asp.

resident number of children between the ages of nine and seventeen is given for each state. Serious emotional disturbances are more often identified in older children and are often identified through the schools and the juvenile justice system. The percentage of children in poverty is provided because poverty has been found to affect the functioning of children and adolescents. Two different functioning scores are provided with both lower and upper limits provided for both. So, for example, the range for Alabama for children and adolescents with serious emotional disturbance would be from a low of 28,567 to a high of 62,848. The definition of the number of children with a serious emotional disturbance is strongly influenced by the criteria used to establish the definition. A definition that accepts a higher level of functioning puts more children and adolescents into the category of seriously emotionally disturbed. Looking at the table for the District of Columbia, which has the highest percentage of children in poverty at 34.8 percent, the estimates of children and adolescents with serious emotional disturbance range from a low of 2,668 to a high of 5,869. Depending on which estimate is accepted, there is a varying impact on policy and the provision of services.

Table 6.3 provides information on the amount of the mental health block grant (MHBG) provided in 2002 to each state mental health agency (SMHA). Each state has a designated SMHA, and the federal government provides annually to each state and the District of Columbia a block grant of funding, which provides flexibility in how each state can use the funding in their state. The table shows that the block grants range in size from a high of $57,240,917 for California to a low of $486,380 for Wyoming.

Federal dollars only account for part of the expenditure for mental health. Table 6.4 shows the state mental health agency expenditures for 2001 both in actual dollars and per capita. The states are also ranked by their total expenditure and then by the amount of money they expend per capita. The table shows that the District of Columbia is ranked number twenty-eight in terms of its expenditure of dollars, but number one in its per capita expenditure of $397.84, the amount that it expends per person. Washington, D.C., is actually an anomaly because the District of Columbia receives much of its budget from Congress. New York, although the second state in size of population, ranks number one in the total dollars it expends on mental health and number

TABLE 6.3
Mental Health Block Grant (MHBG), 2002, State Mental Health Agency (SMHA)

State	Federal mental health services block grant
Alabama	$6,461,856
Alaska	$773,396
Arizona	$6,801,524
Arkansas	$3,704,041
California	$57,240,917
Colorado	$5,195,313
Connecticut	$4,626,918
Delaware	$999,694
Washington, D.C.	$873,273
Florida	$24,966,135
Georgia	$12,328,323
Hawaii	$1,735,391
Idaho	$1,788,346
Illinois	$16,746,583
Indiana	$8,436,498
Iowa	$3,669,110
Kansas	$3,381,059
Kentucky	$5,904,995
Louisiana	$6,231,637
Maine	$1,832,692
Maryland	$8,551,938
Massachusetts	$8,650,294
Michigan	$13,608,011
Minnesota	$5,897,230
Mississippi	$3,998,079
Missouri	$7,099,103
Montana	$1,250,526
Nebraska	$2,042,087
Nevada	$2,853,598
New Hampshire	$1,504,125
New Jersey	$12,420,504
New Mexico	$2,239,672
New York	$29,065,640
North Carolina	$10,239,850
North Dakota	$848,699

continues

TABLE 6.3 *(continued)*
Mental Health Block Grant (MHBG), 2002, State Mental Health Agency (SMHA)

State	Federal mental health services block grant
Ohio	$15,452,581
Oklahoma	$4,687,363
Oregon	$4,333,717
Pennsylvania	$16,837,501
Rhode Island	$1,463,788
South Carolina	$5,648,872
South Dakota	$910,089
Tennessee	$8,217,543
Texas	$30,366,121
Utah	$2,737,465
Vermont	$847,429
Virginia	$10,953,173
Washington	$8,704,091
West Virginia	$2,697,936
Wisconsin	$6,868,644
Wyoming	$486,380

Source: Center for Mental Health Services,
http://mentalhealth.samhsa.gov/cmhs/MentalHealthStatistics/default.asp.

two in its per capita expenditure, which is $175.97, considerably less than Washington, D.C., but a reflection of actual state dollars spent.

Using annual data, comparisons can be made showing how a state's mental health spending is occurring over time. Spending can be compared against inflation to see whether a state is keeping up with inflation. Also improvements or declines in per capita rankings can be identified.

Table 6.5 displays the number of state and county psychiatric hospital inpatient beds that were available in 1998 in the states, the District of Columbia, and Puerto Rico. Over time the number of beds in psychiatric hospitals has declined markedly, as treatment has shifted to the community. Comparing with earlier years,

TABLE 6.4
State Mental Health Agency Actual Dollar and Per Capita Expenditures, 2001

State	Total SMHA expenditures	Total rank	Fiscal year 2001 per capita	Per capita rank
Alabama	$253,279,095	26	$56.97	39
Alaska	$51,444,549	48	$81.36	22
Arizona	$472,341,791	15	$89.36	18
Arkansas	$75,737,397	42	$28.25	50
California	$3,147,792,993	2	$91.61	15
Colorado	$282,614,825	25	$64.24	31
Connecticut	$439,519,867	17	$128.85	6
Delaware	$73,505,846	44	$92.70	14
District of Columbia	$226,558,837	28	$397.84	1
Florida	$578,266,440	12	$35.41	47
Georgia	$380,647,277	21	$45.59	43
Hawaii	$213,643,908	29	$175.21	3
Idaho	$60,524,315	46	$46.01	42
Illinois	$789,861,370	6	$63.54	32
Indiana	$394,000,682	19	$64.70	30
Iowa	$213,046,761	30	$73.18	26
Kansas	$161,844,236	33	$60.31	36
Kentucky	$196,918,103	32	$48.64	41
Louisiana	$200,926,081	31	$45.18	44
Maine	$137,507,731	35	$107.31	11
Maryland	$677,806,345	10	$126.62	7
Massachusetts	$682,218,519	9	$107.38	10
Michigan	$895,065,635	4	$89.96	17
Minnesota	$517,963,917	14	$104.60	12
Mississippi	$246,792,149	27	$86.71	21
Missouri	$336,198,023	23	$59.96	37
Montana	$111,722,233	38	$124.04	8
Nebraska	$86,563,973	40	$50.73	40
Nevada	$120,210,842	37	$57.31	38
New Hampshire	$140,484,321	34	$112.03	9
New Jersey	$763,057,140	7	$90.31	16
New Mexico	$59,378,289	47	$32.60	49
New York	$3,331,688,218	1	$175.97	2
North Carolina	$616,120,223	11	$75.57	24
North Dakota	$49,853,968	49	$78.90	23

continues

TABLE 6.4 *(continued)*
State Mental Health Agency Actual Dollar and Per Capita Expenditures, 2001

State	Total SMHA expenditures	Total rank	Fiscal year 2001 per capita	Per capita rank
Ohio	$692,287,984	8	$61.12	34
Oklahoma	$136,072,416	36	$39.49	45
Oregon	$336,847,640	22	$97.39	13
Pennsylvania	$1,859,763,966	3	$151.98	4
Puerto Rico	$72,184,769	45	$18.88	52
Rhode Island	$92,499,566	39	$87.71	20
South Carolina	$299,401,952	24	$73.99	25
South Dakota	$45,696,380	51	$60.65	35
Tennessee	$395,202,700	18	$69.13	28
Texas	$796,974,433	5	$37.53	46
Utah	$73,790,335	43	$32.64	48
Vermont	$79,658,335	41	$130.46	5
Virginia	$466,472,662	16	$65.18	29
Washington	$525,564,708	13	$88.13	19
West Virginia	$45,804,114	50	$25.52	51
Wisconsin	$389,416,626	20	$72.39	27
Wyoming	$30,097,068	52	$61.12	33

Source: Center for Mental Health Services, http://mentalhealth.samhsa.gov/cmhs/MentalHealthStatistics/default.asp.

one could see the decline in inpatient beds. The state with the largest population, California, had 4,182 psychiatric beds in 1998. The state with the largest number of psychiatric beds was New York, with 7,851. One of the smallest-population states, Alaska, had the least number of beds. Generally states with small populations have smaller numbers of beds and states with large populations have larger numbers of beds, but there is not a direct correlation because there has been some variation in the extent of deinstitutionalization. The number of beds on a per capita basis has not been provided. The inpatient census at the end of 2000 is displayed in Table 6.6. for the states and the District of Columbia. The census covers the number of patients in the state and county

TABLE 6.5
State and County Psychiatric Hospital Inpatient Beds, 1998

State	Number of beds
Alabama	1195
Alaska	79
Arizona	328
Arkansas	182
California	4182
Colorado	698
Connecticut	466
Delaware	334
District of Columbia	813
Florida	2585
Georgia	1255
Hawaii	168
Idaho	196
Illinois	2421
Indiana	1591
Iowa	391
Kansas	592
Kentucky	866
Louisiana	1158
Maine	310
Maryland	1482
Massachusetts	784
Michigan	1863
Minnesota	1161
Mississippi	2635
Missouri	1411
Montana	248
North Carolina	2409
Nebraska	599
Nevada	208
New Hampshire	311
New Jersey	2582
New Mexico	445
New York	7851
North Dakota	283

continues

TABLE 6.5 *(continued)*
State and County Psychiatric Hospital Inpatient Beds, 1998

State	Number of beds
Ohio	1857
Oklahoma	595
Oregon	578
Pennsylvania	4056
Puerto Rico	428
South Carolina	1023
South Dakota	383
Tennessee	1088
Texas	2479
Utah	343
Vermont	80
Virginia	3114
Washington	1185
West Virginia	240
Wisconsin	652
Wyoming	117

Source: Center for Mental Health Services, http://mentalhealth.samhsa.gov/cmhs/MentalHealthStatistics/default.asp.

psychiatric hospitals at the end of the census year. A comparison between the inpatient census and the number of beds available is not possible using these two tables because Table 6.5 is for 1998 data and Table 6.6 is for end-of-2000 data. For example the census for California in 2000 is 4,592, which is actually higher than the number of beds available in 1998, 4,182. Vermont and Alaska had the lowest census at 57 and 59 respectively.

TABLE 6.6
State and County Psychiatric Hospitals, Inpatient Census, End of 2000

State	Census
Alabama	1153
Alaska	59
Arizona	306
Arkansas	158
California	4592
Colorado	788
Connecticut	522
Delaware	342
District of Columbia	656
Florida	2451
Georgia	2916
Hawaii	62
Idaho	163
Illinois	1617
Indiana	1894
Iowa	322
Kansas	459
Kentucky	482
Louisiana	881
Maine	126
Maryland	1430
Massachusetts	642
Michigan	1333
Minnesota	1091
Mississippi	809
Missouri	1408
Montana	171
North Carolina	1989
Nebraska	745
Nevada	317
New Hampshire	173
New Jersey	3154
New Mexico	144
New York	6026
North Dakota	250

continues

TABLE 6.6 *(continued)*
State and County Psychiatric Hospitals, Inpatient Census, End of 2000

State	Census
Ohio	1112
Oklahoma	393
Oregon	536
Pennsylvania	2699
South Carolina	993
South Dakota	260
Tennessee	301
Texas	2481
Utah	1714
Vermont	57
Virginia	1674
Washington	1347
West Virginia	232
Wisconsin	1303
Wyoming	93

Source: Center for Mental Health Services, http://mentalhealth.samhsa.gov/cmhs/MentalHealthStatistics/default.asp.

Department of Health and Human Services, *Healthy People 2010*

The Department of Health and Human Services is another source of statistics on mental health. Although the statistics cover the health field as a whole, there are a number of statistics that apply to mental health. The Surgeon General began establishing goals and objectives for the nation beginning in 1979 with the report, *Healthy People*. This report, like those to follow, was developed through a consultation process and the use of the best scientific knowledge at the time. *Healthy People 2000: National Health Promotion and Disease Prevention Objectives* followed that report. Both of these documents were the basis for the development of state and community public health plans. Currently *Healthy People*

2010 is the guide for health objectives for the first decade of the new century. States, communities, professional organizations, and researchers can use the mental health objectives and statistics in the report to develop programs to improve health and to do research. *Healthy People 2010* is designed to achieve two overarching goals:

> "*Goal 1: Increase Quality and Years of Healthy Life.* The first goal of *Healthy People 2010* is to help individuals of all ages increase life expectancy and improve their quality of life.
>
> *Goal 2: Eliminate Health Disparities.* The second goal of *Healthy People 2010* is to eliminate health disparities among different segments of the population."

Each of the twenty-eight focus area chapters also contains a concise goal statement. This statement frames the overall purpose of the focus area. Chapter 18, "Mental Health and Mental Disorders," sets goals and objectives for mental health. *Healthy People 2010* is available online at http://www.healthypeople.gov/. Chapter 18 of the 2010 conference edition is available at http://www.mentalhealth.samhsa.gov/features/hp2010/goal.asp.

The following are some objectives taken from *Healthy People 2010* that impact the mental health area:

> "6.2 Reduce the proportion of children and adolescents with disabilities who are reported to be sad, unhappy, or depressed.
>
> Target: 17 percent.
>
> Baseline: 31 percent of children and adolescents aged to 11 years with disabilities were reported to be sad, unhappy, or depressed in 1997.
>
> Target setting method: 45 percent improvement (parity with children and adolescents without disabilities in 1997).
>
> Data source: National Health Interview Survey (NHIS), CDC, NCHS."

Table 6.7 gives the 1997 results for children and adolescents aged four to eleven years reporting themselves as sad, unhappy, or depressed. Percentages are given separately for children and adolescents with and without disabilities. The data is further

TABLE 6.7
Children and Adolescents Reported to Be Sad, Unhappy, or Depressed

	Percent reported to be sad, unhappy, or depressed	
Children and adolescents aged 4 to 11 (1997)	**With disabilities**	**Without disabilities***
Total	31	17
Race and ethnicity		
American Indian or Alaskan	DSU	DSU
Asian or Pacific Islander	DSU	13
Asian	DSU	16
Native Hawaiian and other Pacific Islander	DSU	DSU
Black or African American	DSU	16
White	31	17
Hispanic or Latino	32	16
Not Hispanic or Latino	30	17
Black or African American	DSU	17
White	31	18
Gender		
Female	32	16
Male	30	18
Family income level		
Poor	37	20
Near Poor	31	17
Middle/high income	27	17
Geographic location		
Urban	27	17
Rural	39	16

DNA—Data have not been analyzed. DNC—Data are not collected. DSU—Data are statistically unreliable.

*The total represents the target. Data for population groups by race, ethnicity, gender, socioeconomic status, and geographic location are displayed to further characterize the issue.

Source: Healthy People 2010, 2000.

broken down by race and ethnicity. It can be seen that more than twice as many children and adolescents with disabilities (31 percent) report feeling sad, unhappy, or depressed as children and adolescents without disabilities (17 percent). Poor children and adolescents with disabilities have a 37 percent reporting rate, and rural children with disabilities have a 39 percent reporting rate. This data like other data show poorer outcomes for poor children.

> "6.3 Reduce the proportion of adults with disabilities who report feelings such as sadness, unhappiness, or depression that prevent them from being active.
>
> Target: 7 percent.
>
> Baseline: 28 percent of adults aged eighteen years and older with disabilities reported feelings that prevented them from being active in 1997 (age-adjusted to the year 2000 standard population).
>
> Target setting method: 75 percent improvement (parity with adults aged eighteen years and older without disabilities in 1997).
>
> Data source: National Health Interview Survey (NHIS), CDC, NCHS."

Table 6.8 reports the number of adults who report feelings of sadness, unhappiness, or depression that prevent activity for both people with disabilities (28 percent) and for people without disabilities (7 percent). Data is reported by race and ethnicity, gender, income, education, and geographic location. Hispanic or Latino people with disabilities reported the highest level (40 percent) of persons reporting feelings that prevented activity. Family income has a significant impact on reporting of feelings that prevent activity by people with disabilities, with poor people reporting 38 percent, while middle/high income people report only 21 percent.

Chapter 18 focuses on mental health and mental disorders. The goal for the chapter is "Improve mental health and ensure access to appropriate, quality mental health services." Targets and data are provided for both adult and adolescent suicides:

> "18.1 Reduce the suicide rate.
>
> Target: 5.0 suicides per 100,000 population.

TABLE 6.8
Adults Who Report Feelings that Prevent Activity

	Percent who report feelings that prevent activity	
Adults aged 18 and older (1997)	**With disabilities**	**Without disabilities***
Total	28	7
Race and ethnicity		
American Indian or Alaska	22	15
Asian or Pacific Islander	30	7
Asian	DSU	6
Native Hawaiian and other Pacific Islander	DSU	14
Black or African American	31	8
White	28	7
Hispanic or Latino	40	9
Not Hispanic or Latino	27	7
Black or African American	31	8
White	27	6
Gender		
Female	30	8
Male	26	6
Family income level		
Poor	38	13
Near Poor	30	10
Middle/high income	21	6
Geographic location		
Education level (aged 25 years and older)		
Less than high school	34	10
High school graduate	29	7
At least some college	25	5
Urban	29	7
Rural	26	6

DNA—Data have not been analyzed. DNC—Data are not collected. DSU—Data are statistically unreliable.

Note: Age adjusted to the year 2000 standard population.

*The total represents the target. Data for population groups by race, ethnicity, gender, socioeconomic status, and geographic location are displayed to further characterize the issue.

Source: Healthy People 2010, 2000.

Baseline: 11.3 suicides per 100,000 population occurred in 1998 (age-adjusted to the year 2000 standard population).

Target setting method: Better than the best.

Data source: National Vital Statistics System (NVSS), CDC, NCHS."

"18.2 Reduce the rate of suicide attempts by adolescents.

Target: 12-month average of 1 percent.

Baseline: 12-month average of 2.6 percent of adolescents in grades 9 through 12 attempted suicide in 1999.

Target setting method: Better than the best.

Data source: Youth Risk Behavior Surveillance System (YRBSS), CDC, NCCDPHP."

Table 6.9 reports suicide statistics for adults in 1998 as a rate per 100,000 population. The adult suicide rates are broken down by race and ethnicity, gender, education level, and age. The total suicide rate is 11.3 persons per 100,000 people. The data show that males have a much higher suicide rate at 19.2 than females at 4.3. This difference is in accord with the fact that males are more likely to complete a suicide act, while more females who attempt suicide fail to complete the act. The data also show the suicide rate drops for people with a college education. This data can be tracked over time to determine the success or failure of policies to affect the suicide rate. Table 6.10 shows suicide attempts by adolescents as 2.6 percent. The suicide attempt rate is slightly higher for females than males. Males' attempts and completions of suicide increase with age.

"18.3 Reduce the proportion of homeless adults who have serious mental illness (SMI).

Target: 19 percent.

Baseline: 25 percent of homeless adults aged eighteen years and older had SMI in 1996.

Target setting method: 24 percent improvement. (Better than the best will be used when data are available.)

TABLE 6.9
Rate of Suicides, 1998

Total population, 1998	Suicides (rate per 100,000)
Total	11.3
Race and ethnicity	
American Indian or Alaska	12.6
Asian or Pacific Islander	6.6
Asian	DNC
Native Hawaiian and other Pacific Islander	DNC
Black or African American	5.8
White	12.2
Hispanic or Latino	6.3
Not Hispanic or Latino	11.8
Black or African American	6.0
White	12.8
Gender	
Female	4.3
Male	19.2
Education level (aged 25 to 64)	
Less than high school	17.9
High school graduate	19.2
At least some college	10.0
Age (not age adjusted)	
10 to 14	1.6
15 to 19	8.9
20 to 24	13.6

DNA—Data have not been analyzed. DNC—Data are not collected. DSU—Data are statistically unreliable.

Note: Age adjusted to the year 2000 standard population.

Source: Healthy People 2010, 2000.

Data source: Projects for Assistance in Transition from Homelessness (PATH) Annual Application, SAMHSA, CMHS."

"18.4 Increase the proportion of persons with serious mental illness (SMI) who are employed.

Target: 51 percent.

TABLE 6.10
Suicide Attempts by Adolescents, 1999

Students in Grades 9 through 12, 1999	Suicide attempts (percent)
Total	2.6
Race and ethnicity	
American Indian or Alaska	DSU
Asian or Pacific Islander	DSU
Asian	DSU
Native Hawaiian and other Pacific Islander	DSU
Black or African American	3.1
White	2.2
Hispanic or Latino	3.0
Not Hispanic or Latino	2.6
Black or African American	2.9
White	1.9
Gender	
Female	3.1
Male	2.1
Parents' education level	
Less than high school	DNC
High school graduate	DNC
At least some college	DNC
Sexual orientation	DNC

DNA—Data have not been analyzed. DNC—Data are not collected. DSU—Data are statistically unreliable.

Source: Healthy People 2010, 2000.

Baseline: 43 percent of persons aged eighteen years and older with SMI were employed in 1994.

Target setting method: 19 percent improvement. (Better than the best will be used when data are available.)

Data source: National Health Interview Survey (NHIS), CDC, NCHS."

"18.9 Increase the proportion of adults with mental disorders who receive treatment."

Table 6.11 shows four objectives for increasing the proportion of adults with mental disorders who receive treatment. The objectives are for adults aged eighteen to fifty-four with serious mental illness and for adults aged eighteen and older with

TABLE 6.11
Proportion of Adults with Mental Disorders Who Receive Treatment

Target and baseline			
Objective	**Increase in adults with mental disorders receiving treatment**	**1997 baseline (unless noted) (percent)**	**2010 target (percent)**
18.9a	Adults aged 18 to 54 years with serious mental illness	47 (1991)	55
18.9b	Adults aged 18 years and older with recognized depression	23	50
18.9c	Adults aged 18 years and older with schizophrenia	60 (1984)	75
18.9d	Adults aged 18 years and older with generalized anxiety disorder	38	50

Target setting method: 17 percent improvement for 18.9a. (Better than the best will be used when data are available.) Better than the best for 18.9b, 18.9c, and 18.9d.

Data sources: Epidemiologic Catchment Area (ECA) Program, NIH, NIMH; National Household Survey on Drug Abuse (NHSDA), SAMHSA, OAS; National Comorbidity Survey, SAMHSA, CMHS; NIH, NIMH.

depression, schizophrenia and generalized anxiety disorder. The baseline data are for 1997, but for two objectives the data are older—1991 and 1984. The targets for 2010 are from 50 to 75 percent. Table 6.12 presents the data for 1997 for adults with mental disorders receiving treatment by race and ethnicity.

> "18.12 Increase the number of States and the District of Columbia that track consumers' satisfaction with mental health services they receive.
>
> Target: 50 states and the District of Columbia.
>
> Baseline: 36 states tracked consumers' satisfaction with the mental health services they received in 1999.
>
> Target setting method: Total coverage.
>
> Data source: Mental Health Statistics Improvement Program, SAMHSA."

> "18.14 Increase the number of states, territories, and the District of Columbia with an operational mental health plan

TABLE 6.12
Adults with Mental Disorders Receiving Treatment by Race and Ethnicity

	Received treatment (percent)			
Adults aged 18 years and older with mental disorders, 1997 (unless noted)	**18.9a Serious mental illness (aged 19 to 54 years) (1991)**	**18.9b Recognized depression**	**18.9c Schizophrenia (1984)**	**18.9d Generalized anxiety disorder**
Total	47	23	60	38
American Indian or Alaska Native	DNA	DSU	DSU	DSU
Asian or Pacific Islander	DNA	DSU	DSU	DSU
Asian	DNA	DNC	DSU	DNC
Native Hawaiian and Other Pacific Islander	DNA	DNC	DSU	DNC
Black or African American	DNA	16	DNC	26
White	DNA	24	DNC	39
Hispanic or Latino	DNA	20	42	DSU
Not Hispanic or Latino	DNA	DNC	DNC	40
Black or African American	DNA	DNA	63	DNA
Gender				
Female	DNA	24	63	32
Male	DNA	21	51	49
Education level				
Less than high school	DNA	22	48	48
High school graduate	DNA	22	48	48
At least some college	DNA	28	66	32
Sexual orientation	DNC	DNC	DNC	DNC

DNA—Data have not been analyzed. DNC—Data are not collected. DSU—Data are statistically unreliable.

that addresses mental health crisis interventions, ongoing screening, and treatment services for elderly persons.

Target: Fifty states and the District of Columbia.

Baseline: Twenty-four states had an operational mental health plan that addressed mental health crisis interventions, ongoing screening, and treatment services for elderly persons in 1997.

Target setting method: Total coverage.

Data source: National Technical Assistance Center for State Mental Health Systems, National Association of State Mental Health Program Directors, National Research Institute; SAMHSA, CMHS." (*Healthy People 2010*, 2000).

National Center for Health Statistics, Mental Health Disorders Statistics

Statistics for mental health disorders are provided by the National Center for Health Statistics under FASTSTATS A to Z at http://www.cdc.gov/nchs/fastats/mental.htm. These statistics are morbidity statistics, that is, statistics covering the number of people ill with a certain disease or disorder and also health care–use statistics covering the number of people with a disorder seen at a certain level of care. Also provided are mortality data for suicides. These statistics are taken from various sources and represent data from various years. One needs to be careful in using the data to pay close attention to the year of the data. Also one can note the sources of the data and go to the original source for further information.

Morbidity

"Percent of noninstitutionalized adults with severe psychological distress in the past 30 days: 3.1 (2003)."

Source: *Early Release of Selected Estimates Based on Data from the January-September 2004 National Health Interview Survey* (2004), http:www.cdc.gov/nchs/about/major/nhis/released100503.htm.

Ambulatory Care

"Number of visits to office-based physicians for mental disorders: 40 million (2002)."

Source: *National Ambulatory Medical Care Survey: 2002 Summary* (2002), http:/www.cdc.gov/nchs/data/ad/ad346.pdf.

"Number of hospital emergency department visits for mental disorders: 3.7 million (2003)."

Source: *National Hospital Ambulatory Medical Care Survey: 2003 Emergency Department Summary* (2003), http://www.cdc.gov/nchs/data/ad/ad358.pdf.

"Number of ambulatory care visits for mental disorders: 49.2 million (1997).

Number of ambulatory care visits for depression: 19.4 million (1997).

Number of ambulatory care visits for schizophrenia and other psychoses: 7.7 million (1997).

Number of ambulatory care visits for anxiety: 5.7 million (1997).

Number of ambulatory care visits related to drugs or alcohol: 3.6 million (1997).

Number of ambulatory care visits for attention deficit disorder: 3.5 million (1997)."

Source: *Ambulatory Care Visits to Physician Offices, Hospital Outpatient Departments, and Emergency Departments* (1997), http://www.cdc.gov/nchs/data.

Hospital Inpatient Care

"Number of discharges for mental disorders: 2.5 million (2002).

Average length of stay for mental disorders: 7.1 days (2002)."

Source: *2002 National Hospital Discharge Survey* (2002), http://www.cdc.gov./nchs/data/ad/ad342.pdf.

Nursing Home Care

"Number of residents with mental disorders, including Alzheimer's disease, as primary diagnosis: 445,800 (1999).

Percent of residents with mental disorders, including Alzheimer's disease, as primary diagnosis: 27.4 (1999).

Number of residents with Alzheimer's disease and other dementias as primary diagnosis: 207,000 (1999)."

Source: *National Nursing Home Survey: 1999 Summary* (1999), http://www.cdc.gov/nchs/data/series/sr_13/sr_152.pdf.

Mortality

"Number of suicide deaths: 31,655 (2002).

Suicide deaths per 100,000 resident population: 11.0 (2002)."

Source: *Deaths: Final Data for 2002*. National Center for Health Statistics (2005), http://www.cdc.gov/nchs/data/nvsr/nvsr52/nvsr52_03.pdf.

National Institute of Mental Health

As the lead federal agency for research on mental and behavioral disorders, the National Institute of Mental Health (NIMH) makes available fact sheets and other information, including statistics on mental disorders, on their Web site. Its mission is "to reduce the burden of mental illness and behavioral disorders through research on mind, brain, and behavior." It is mandated to focus on the people with the most serious mental illness, and this mandate is reflected in its expenditures, as shown in Table 6.13. The National Institute of Mental Health can be reached at http://www.nimh.nih.gov/.

NIMH provides booklets and fact sheets with the latest research-based information on mental illness. NIMH conducts public education campaigns to educate the public about mental disorders. One of its recent campaigns, "Real Men Real Depression," is aimed at increasing the recognition of depressive disorders in men, who are often underdiagnosed and undertreated for

TABLE 6.13
National Institute of Mental Health Research Funding

	NIMH Research Funding		
	Fiscal Year 1999 (dollars)	Fiscal Year 2003 (dollars)	Increase over Fiscal Year 1999
Major Depression	112,654,749	212,766,139	89%
Schizophrenia	196,515,314	311,812,660	59%
Bipolar Disorder	57,805,403	86,882,796	50%
Autism	19,278,940	51,096,483	165%
All NIMH Research	823,528,000	1,285,475,000	56%

Source: National Institute of Mental Health, 2003.

depression. Its educational materials are a source of data on various mental disorders. Its Web site provides information on mental disorders affecting children and adults. It has information available on "depression, bipolar disorder, schizophrenia, anxiety disorders, eating disorders, suicide that occurs in the context of mental disorders, autism-spectrum disorders, attention deficit hyperactivity disorders, and other behavioral conditions that can adversely affect a child's healthy development." There are also consensus conference reports, patient education materials, and links to other federal government Web sites and resources including the *Surgeon General's Report on Mental Health,* and information on the report *America's Children: Key National Indicators of Well-Being 2005,* which shows that nearly 5 percent of children, 2.7 million, are reported by their parents to suffer from definite or severe behavioral or emotional disorders that may interfere with their ability to learn, friendships, and family life.

The Impact of Mental Illness on Society—World Health Organization, *Global Burden of Disease*

There has long been a failure to recognize the significance of the impact of mental disabilities in the United States and the world. The World Health Organization (WHO), along with the World Bank and Harvard University, published a report, *Global Burden of Disease,* that shows that mental illness, including suicide, accounts for over 15 percent of the disease burden in developed economies including the United States. This is a greater disease burden than cancer. Disease conditions were compared using a measure called the disability-adjusted life years (DALY). This measure was developed to measure lost years of healthy life. Years could be lost to either disability or premature death. The measure was also weighted for the severity of the disability. Major depression was found to be equal to paraplegia or blindness in degree of disability, while schizophrenia was found to be equivalent to quadriplegia in degree of disability. Ischemic heart disease was ranked number one in magnitude of disease burden in developed countries, while major depression ranked second.

The *Global Burden of Disease* Web site can be found at www.who.int/msa/mnh/ems/dalys/intro.htm.

The world's population is aging, particularly in developed countries, and even in developing countries infectious diseases are being replaced by chronic disease and disability. Schizophrenia, bipolar disorder, obsessive-compulsive disorder, panic disorder, and post-traumatic stress disorder, which now all contribute significantly to the total burden of illness, can expect to have even a greater impact in the future.

The following statistics come from the *Global Burden of Disease* (1996):

> Major depression is the leading cause of *disability* (measured by the number of years *lived* with a disabling condition) worldwide among persons age 5 and older. For women throughout the world as well as those in established market economies, depression is the leading cause of DALYs. In established market economies, schizophrenia and bipolar disorder are also among the top 10 causes of DALYs for women.

Further statistics can be found in Murray and Lopez, eds., *The Global Burden of Disease and Injury Series, Volume 1: A Comprehensive Assessment of Mortality and Disability from Diseases, Injuries, and Risk Factors in 1990 and Projected to 2020.* Cambridge, MA: Published by the Harvard School of Public Health on behalf of the World Health Organization and the World Bank, Harvard University Press, 1996.

Table 6.14 shows the leading sources of disease burden in established market economies for 1990. Of the top eight causes, unipolar major depression ranks second.

Table 6.15 shows the disease burden by selected illness categories in established market economies in 1990. Of seven illness categories, three involve mental and behavioral disorders including suicide which ranks second, alcohol use which ranks fifth, and drug use which ranks seventh. Table 6.16 shows mental illness as a source of disease burden in established market economies in 1990. The top mental illnesses are unipolar major depression, schizophrenia, bipolar disorder, obsessive-compulsive disorder, panic disorder, post-traumatic stress disorder, and suicide.

TABLE 6.14
Leading Sources of Disease Burden in Established Market Economies, 1990

	Total (millions)*	Percent of total
All causes	98.7	
Ischemic heart disease	8.9	9.0
Unipolar major depression	6.7	6.8
Cardiovascular disease	5.0	5.0
Alcohol use	4.7	4.7
Road traffic accidents	4.3	4.4
Lung & UR cancers	3.0	3.0
Dementia & degenerative CNS	2.9	2.9
Osteoarthritis	2.7	2.9
Diabetes	2.4	2.4
COPD	2.3	2.3

*Measured in DALYs (lost years of healthy life regardless of whether the years were lost to premature death or disability).

Source: World Health Organization, 1996.

TABLE 6.15
Disease Burden by Selected Illness Categories in Established Market Economies, 1990

All cardiovascular conditions	18.6
All mental illness including suicide	15.4
All malignant disease (cancer)	15.0
All respiratory conditions	4.8
All alcohol use	4.7
All infectious and parasitic disease	2.8
All drug use	1.5

Source: World Health Organization, 1996.

TABLE 6.16
Mental Illness as a Source of Disease Burden in Established Market Economies, 1990

	Total (millions)*	Percent of total
All causes	98.7	
Unipolar major depression	6.7	6.8
Schizophrenia	2.3	2.3
Bipolar disorder	1.7	1.7
Obsessive-compulsive disorder	1.5	1.5
Panic disorder	0.7	0.7
Post-traumatic stress disorder	0.3	0.3
Self-inflicted injuries (suicide)	2.2	2.2
All mental disorders	15.3	15.4

*Measured in DALYs (lost years of healthy life regardless of whether the years were lost to premature death or disability).

Source: World Health Organization, 1996.

Mental Disorders in America

In any year, about 22 percent of adult Americans have a diagnosable mental disorder (Regier, Narrow, Rae, et al. 1993). When applied to the 2005 estimated U.S. population of 275,633,416, this means there are approximately 55.3 million people with mental disorders. The World Health Organization's *Global Burden of Disease* has found that four of the ten leading causes of disability in the United States and other developed countries are mental disorders—major depression, bipolar disorder, schizophrenia, and obsessive-compulsive disorder (Murray and Lopez 1996). Also some people have more than one mental disorder. The National Institute of Mental Health, a part of the National Institutes of Health in the Department of Health and Human Services, makes available fact sheets on mental health. NIH Publication No. 01–4584 makes available the statistics that follow under each disorder type.

Depressive Disorders

Depressive disorders include major depressive disorder, bipolar disorder, and dysthymic disorders.

- "Approximately 18.8 million American adults, (Narrow, 1998) or about 9.5 percent of the U.S. population age 18 and older in a given year (Regier et al., 1993) have a depressive disorder."
- "Nearly twice as many women (12.0 percent) as men (6.6 percent) are affected by a depressive disorder each year. These figures translate to 12.4 million women and 6.4 million men in the U.S. (Narrow, 1998)."
- "Depressive disorders may be appearing earlier in life in people born in recent decades compared to the past. (Weissman, Bland, Canino, et al, 1999)."
- "Depressive disorders often co-occur with anxiety disorders and substance abuse. (Regier, et al, 1998)."

In the United States and developed countries worldwide, major depressive disorder is the leading cause of disability (Murray and Lopez 1996).

- "Major depressive disorder affects approximately 9.9 million American adults (Narrow, 1998), or about 5.0 percent of the U.S. population age 18 and older in a given year (Regier, et al, 1993)."
- "Nearly twice as many women (6.5 percent) as men (3.3 percent) suffer from major depressive disorder each year. These figures translate to 6.7 million women and 3.2 million men (Narrow, 1998)."
- "While major depressive disorder can develop at any age, the average age at onset is the mid-twenties (American Psychiatric Association, 1994)."

Dysthymic disorder is a chronic mild depression that lasts at least two years in adults and one year in children.

- "Dysthymic disorder affects approximately 5.4 percent of the U.S. population age 18 and older during their lifetime. (Regier, 1993) This figure translates to about 10.9 million American adults (Narrow, 1998)."

- "About 40 percent of adults with dysthymic disorder also meet criteria for major depressive disorder or bipolar disorder in a given year (Regier, 1993)."
- "Dysthymic disorder often begins in childhood, adolescence, or early adulthood (American Psychiatric Association, 1994)."

Bipolar disorder consists of episodes of major depression followed by manic episodes of speeded-up emotions and responses.

- "Bipolar disorder affects approximately 2.3 million American adults, (Narrow, 1998) or about 1.2 percent of the U.S. population age 18 and older in a given year (Regier, et al, 1993)."
- "Men and women are equally likely to develop bipolar disorder (Narrow, 1998)."
- The average age at onset for a first manic episode is the early twenties (American Psychiatric Association, 1994)."

Suicide

Suicide may occur without mental illness, or suicide can be the outcome of mental illness, particularly major depression.

- "In 2000, 29,350 people died by suicide in the U.S. (Minino, et al, 2002)."
- "More than 90 percent of people who kill themselves have a diagnosable mental disorder, commonly a depressive disorder or a substance abuse disorder. (Conwell and Brent, 1995)."
- "The highest suicide rates in the U.S. are found in white men over age 85 (Minino, et al, 2002)."
- "In 2000, suicide was the third leading cause of death among 15 to 24-year-olds (Minino, et al, 2002)"
- "Four times as many men as women die by suicide (Minino, et al, 2002); however, women attempt suicide two to three times as often as men (Weissman, et al)."

The National Institutes of Health (2005) also supply a number of facts and statistics on suicide in the United States. The following statistics are for 2001. They show the seriousness of suicide in the United States.

- "Suicide was the 11th leading cause of death in the United States.
- It was the eighth leading cause of death for males, and 19th leading cause of death for females.
- The total number of suicide deaths was 30,622.
- The 2001 age-adjusted rate** was 10.7/100,000 or 0.01 percent.
- 1.3 percent of total deaths were from suicide. By contrast, 29 percent were from diseases of the heart, 23 percent were from malignant neoplasms (cancer), and 6.8 percent were from cerebrovascular disease (stroke), the three leading causes.
- Suicides outnumbered homicides (20,308) by three to two.
- There were twice as many deaths due to suicide than deaths due to HIV/AIDS (14,175).
- Suicide by firearms was the most common method for both men and women, accounting for 55 percent of all suicides.
- More men than women die by suicide.
- The gender ratio is 4:1.
- 73 percent of all suicide deaths are white males.
- 80 percent of all firearm suicide deaths are white males.
- Among the highest rates (when categorized by gender and race) are suicide deaths for white men over 85, who had a rate of 54/100,000.
- Suicide was the third leading cause of death among young people 15 to 24 years of age, following unintentional injuries and homicide. The rate was 9.9/100,000 or .01 percent.
- The suicide rate among children ages 10–14 was 1.3/100,000 or 272 deaths among 20,910,440 children in this age group. The gender ratio for this age group was 3:1 (males: females).
- The suicide rate among adolescents aged 15–19 was 7.9/100,000 or 1,611 deaths among 20,271,312 adolescents in this age group. The gender ratio for this age group was 5:1 (males: females).
- Among young people 20 to 24 years of age, the suicide rate was 12/100,000 or 2,360 deaths among 19,711,423

people in this age group. The gender ratio for this age group was 7:1 (males: females).
- No annual national data on all attempted suicides are available.
- Other research indicates that: there are an estimated 8–25 attempted suicides for each suicide death; the ratio is higher in women and youth and lower in men and the elderly.
- More women than men report a history of attempted suicide, with a gender ratio of 3:1."

"*2001 U.S. mortality data are based on the International Classification of Disease, 10th revision (ICD-10), whereas ICD-9 has been used from 1979–1998. For this reason, comparisons between data from years 1999–2001 and earlier mortality data should be made carefully. For a full explanation of the implications of this change, see http://www.cdc.gov/ncipc/wisqars/fatal/help/datasources.htm#6.3 (NIMH, 2005)."

"**Age-adjusted rates refer to weighting rates by a population standard to allow for comparisons across time and among risk groups. The 2001 mortality data are calculated using figures from the 2000 census, whereas previous years have been calculated using 1940 census data. For this reason, comparisons between data from years 2000 to 2001 and earlier mortality data should be made carefully. For a full explanation of the implications of this change, see http://www.cdc.gov/ncipc/wisqars/fatal/help/datasources.htm#6.2."

Schizophrenia

Schizophrenia is a complex disease that through further research may be found to be several different diseases.

- "Approximately 2.2 million American adults, (Narrow, 1998) or about 1.1 percent of the population age 18 and older in a given year, (Regier, 1993) have schizophrenia."
- "Schizophrenia affects men and women with equal frequency (Robins and Regier, 1991)."
- "Schizophrenia often first appears earlier in men, usually in their late teens or early twenties, than in women, who are generally affected in their twenties or early thirties (Robins and Regier, 1991)."

Anxiety Disorders

The category of anxiety disorders under the DSM-IV includes generalized anxiety disorder, panic disorder, obsessive-compulsive disorder, post-traumatic stress disorder, and phobias.

"Approximately 19.1 million American adults ages eighteen to fifty-four, or about 13.3 percent of people in this age group in a given year, have an anxiety disorder. (Narrow, Rae, and Regier, 1998)."

- "Anxiety disorders frequently co-occur with depressive disorders, eating disorders, or substance abuse (Regier, 1998; Wonderlich and Mitchell, 1997)."
- "Many people have more than one anxiety disorder (Robins, and Regier, 1991)."
- "Women are more likely than men to have an anxiety disorder. Approximately twice as many women as men suffer from panic disorder, post-traumatic stress disorder, generalized anxiety disorder, agoraphobia, and specific phobia, though about equal numbers of women and men have obsessive-compulsive disorder and social phobia (Bourdon, et al., 1998; Davidson, 2000; Robins and Regier, 1991)."
- "Approximately 2.4 million American adults ages 18 to 54, or about 1.7 percent of people in this age group in a given year, have panic disorder (Narrow, Rae, and Regier, 1998)."
- "Panic disorder typically develops in late adolescence or early adulthood (Robins and Regier, 1991)."
- "About one in three people with panic disorder develop *agoraphobia*, a condition in which they become afraid of being in any place or situation where escape might be difficult or help unavailable in the event of a panic attack (Robins and Regier, 1991)."
- "Approximately 3.3 million American adults ages 18 to 54, or about 2.3 percent of people in this age group in a given year, have OCD (obsessive-compulsive disorder) (Narrow, Rae, and Regier, 1998)."
- "The first symptoms of OCD often begin during childhood or adolescence (Robins and Regier, 1991)."

- "Approximately 5.2 million American adults ages 18 to 54, or about 3.6 percent of people in this age group in a given year, have PTSD (post-traumatic stress disorder) (Narrow, Rae, and Regier, 1998)."
- "PTSD can develop at any age, including childhood (American Academy of Child and Adolescent Psychiatry, 1998)."
- "About 30 percent of Vietnam veterans experienced PTSD at some point after the war (Kulka, et al., 1988). The disorder also frequently occurs after violent personal assaults such as rape, mugging, or domestic violence; terrorism; natural or human-caused disasters; and accidents."
- "Approximately 4.0 million American adults ages 18 to 54, or about 2.8 percent of people in this age group in a given year, have GAD (generalized anxiety disorder) (Narrow, Rae, and Regier)."
- "GAD can begin across the life cycle, though the risk is highest between childhood and middle age (Robbins and Regier, 1991)."
- "Approximately 5.3 million American adults ages 18 to 54, or about 3.7 percent of people in this age group in a given year, have social phobia (Narrow, Rae, and Regier)."
- "Social phobia typically begins in childhood or adolescence (Robbins and Regier, 1991)."
- "*Agoraphobia* involves intense fear and avoidance of any place or situation where escape might be difficult or help unavailable in the event of developing sudden panic-like symptoms. Approximately 3.2 million American adults ages 18 to 54, or about 2.2 percent of people in this age group in a given year, have agoraphobia (Narrow, Rae, and Regier, 1998)."
- "*Specific phobia* involves marked and persistent fear and avoidance of a specific object or situation. Approximately 6.3 million American adults ages 18 to 54, or about 4.4 percent of people in this age group in a given year, have some type of specific phobia (Narrow, Rae, and Regier, 1998)."

Eating Disorders

Eating disorders occur usually as binge-eating disorder, anorexia nervosa, and bulimia nervosa.

- "Females are much more likely than males to develop an eating disorder. Only an estimated 5 to 15 percent of people with anorexia or bulimia (Andersen, 1995) and an estimated 35 percent of those with binge-eating disorder (Spitzer, et al., 1993) are male."
- "In their lifetime, an estimated 0.5 percent to 3.7 percent of females suffer from anorexia and an estimated 1.1 percent to 4.2 percent suffer from bulimia (American Psychiatric Association Work Group on Eating Disorders, 2000)."
- "Community surveys have estimated that between 2 percent and 5 percent of Americans experience binge-eating disorder in a 6-month period (Bruce and Agras, 1992; Spitzer, et al, 1993)."
- "The mortality rate among people with anorexia has been estimated at 0.56 percent per year, or approximately 5.6 percent per decade, which is about 12 times higher than the annual death rate due to all causes of death among females ages 15–24 in the general population (Sullivan, 1995)."

Attention Deficit Hyperactivity Disorder (ADHD)

ADHD is one of the most common mental disorders in children and adolescents.

- "It affects an estimated 4.1 percent of youths ages 9 to 17 in a 6-month period (Shaffer et al., 1996)."
- "About two to three times more boys than girls are affected (Wolraich, Hannah, Baumgaertel, and Feurer, 1998)."
- "ADHD usually becomes evident in preschool or early elementary years. The disorder frequently persists into adolescence and occasionally into adulthood (Barkley, 1996)."

Department of Health and Human Services, Substance Abuse and Mental Health Services Administration

In the Department of Health and Human Services (DHHS) is the Substance Abuse and Mental Health Services Administration (SAMHSA). In 1992 Public Law 102-321 established SAMHSA to create programs and provide funding to address the problems of mental and substance abuse disorders in the United States. SAMHSA's mission and vision are now focused on President Bush's administration priorities. The vision is consistent with the President's New Freedom Initiative, which promotes a life in the community for everyone. SAMHSA's mission is "to build resilience and facilitate recovery for people with or at risk for substance abuse and mental illness." Table 6.17 provides information on the expenditure of the 2005 budget for SAMHSA. The table shows expenditures for mental illness and alcohol and substance abuse treatment. The Office of Applied Studies provides statistics on drug abuse and an online database archive. SAMHSA can be found at http://www.samhsa.gov/Menu/Level2_about.aspx.

TABLE 6.17

Substance Abuse and Mental Health Services Administration Funding by Program Priority Area (Dollars by thousands)

			Fiscal year 2005	Proposal
Program priority area	Fiscal year 2003	Fiscal year 2004	Estimated amount	+/− fiscal year 2004
CMHS	4,668	9,200	15,200	+6,000
CSAP	0	0	0	0
CSAT	6,689	6,610	6,610	0

continues

TABLE 6.17 *(continued)*
Substance Abuse and Mental Health Services Administration Funding by Program Priority Area (Dollars by thousands)

			Fiscal year 2005	Proposal
Program priority area	**Fiscal year 2003**	**Fiscal year 2004**	**Estimated amount**	**+/− fiscal year 2004**
Co-occurring disorders	11,357	15,810	21,810	+6,000
CMHS	0	0	0	
CSAP	0	0	0	
CSAT	162,021	255,693	355,167	+99,474
Block grant	1,403,146	1,423,317	1,465,788	+42,471
Substance abuse treatment capacity	1,565,167	1,679,010	1,820,955	+141,945
CMHS	1,845	2,500	2,500	
CSAP	0	0	0	
CSAT	0	0	0	
Seclusion and restraint	1,845	2,500	2,500	
CMHS	14,662	9,503	7,332	−2,171
CSAP	146,485	156,184	156,391	+207
CSAT	350,786	355,829	366,447	+10,618
Strategic prevention framework[b]	511,933	521,516	530,170	+8,654
CMHS				
PRNS	134,384	134,460	138,198	+3,738
Children's M/H services	98,053	102,353	106,013	+3,660
CSAP	9,743	1,653	0	−1,653
CSAT	24,621	33,700	33,700	−0
Children and families	266,801	272,166	277,911	+5,745
CMHS	44,245	47,126	80,483	+33,357
Protection and advocacy	33,779	34,620	34,620	
M/H block grant	437,140	434,690	436,070	+1,380
CSAP	90	63	63	
CSAT	540	537	537	+0
Mental health system transformation	515,794	517,036	551,773	+34,737
CMHS	9,134	3,623	389	−3,234
CSAP	994	994	0	−994
CSAT	1,696	1,686	0	−1,686

continues

TABLE 6.17 *(continued)*
Substance Abuse and Mental Health Services Administration Funding by Program Priority Area (Dollars by thousands)

Program priority area[a]	Fiscal year 2003	Fiscal year 2004	Fiscal year 2005 Estimated amount	Proposal +/− fiscal year 2004
Disaster readiness and response	11,824	6,303	389	–5,914
CMHS	12,090	12,019	12,019	
PATH	43,073	49,760	55,251	+5,491
Samaritan initiative	0	0	10,000	+10,000
CSAP	0	0	0	
CSAT	34,153	33,952	33,977	+25
Homelessness	89,316	95,731	111,247	+15,516
CMHS	4,960	4,970	0	–4,970
CSAP	0	0	0	0
CSAT	0	0	0	0
Aging	4,960	4,070	---	–4,970
CMHS	10,498	10,436	10,492	+56
CSAP	39,799	39,564	39,564	
CSAT	61,807	61,442	61,442	+0
HIV/AIDS and Hepatitis[c]	112,104	111,442	111,498	+56
CMHS	7,957	6,959	3,935	–3,024
CSAP	0	0	0	
CSAT	25,751	25,599	25,599	–0
Criminal justice	33,708	32,558	29,534	–3,024
CMHS	856,488	862,219	912,502	+50,283
CSAP	197,111	198,458	196,018	–2,440
CSAT	2,071,210	2,198,365	2,349,267	+150,902
Total[d]	3,124,809	3,259,042	3,457,787	+198,745

CMHS—Mental health services. CSAP—Substance abuse prevention. CSAT—Substance abuse treatment.

[a] Represents primary program category; may relate to other categories: reflects comparable adjustments for prevention/early intervention; change to Strategic Prevention Framework and New Freedom Initiative; change to Mental Health System Transformation.

[b] Includes 20 percent prevention set-aside from SAPTBG.

[c] Excludes HIV/AIDS set-aside from SAPTBG.

[d] Excludes all program management funds including PHS evaluation. Includes PHS evaluation funds applicable to PRNS and the SAPT Block Grant. Excludes St. Elizabeth's funding for fiscal year 2003.

Source: SAMHSA, 2005, http://www.samhsa.gov/budget/B2005/spending/cj 12.aspx.

Nonprofit Organizations and Mental Health Statistics

Nonprofit organizations are also a source for statistics on mental health. The National Mental Health Association, for example, provides a fact sheet on the prevalence and economic costs of mental illness. Many of the statistics on the fact sheet come from the federal government, but the fact sheet covers a wide array of statistics regarding the general public, minorities, and families. The National Mental Health Association Web site can be found at http://www.nmha.org/ and the following information comes from their Web site.

General Public

- "More than 54 million Americans have a mental disorder in any given year, although fewer than 8 million seek treatment (Substance Abuse and Mental Health Services Administration (SGRMH), 1999)."
- "Depression and anxiety disorders—the two most common mental illnesses—each affect 19 million American adults annually (National Institute of Mental Health (NIMH), 1999)."
- "Approximately 12 million women in the United States experience depression every year—roughly twice the rate of men (NIMH, 1999)."
- "One percent of the population (more than 2.5 million Americans) has schizophrenia (Schizophrenia Bulletin, 1998)."
- "Bipolar disorder, also known as manic-depressive illness, affects more than 2 million Americans (NIMH, 2000)."
- "Each year, eating disorders such as anorexia nervosa and bulimia nervosa affect millions of Americans, 85–90 percent of whom are teens and young adult women (National Mental Health Association (NMHA), 2000)."
- "Depression greatly increases the risk of developing heart disease. People with depression are four times more likely to have a heart attack than those with no history of depression (NIMH, 1998)."

- "Approximately 15 percent of all adults who have a mental illness in any given year also experience a co-occurring substance abuse disorder, which complicates treatment (Surgeon General's Report on Mental Health (SGRMH), 1999)."
- "Up to one-half of all visits to primary care physicians are due to conditions that are caused or exacerbated by mental or emotional problems (Collaborative Family Healthcare Coalition (CFHC), 1998)."

Minorities

- "Adult Caucasians who have either depression or an anxiety disorder are more likely to receive treatment than adult African Americans with the same disorders even though the disorders occur in both groups at about the same rate, taking into account socioeconomic factors (SGRMH, 1999)."
- "More than half of all African Americans and Native Americans are anticipated to use public insurance to pay for inpatient mental health treatment, compared to 34 percent of Caucasians (SAMHSA, 1998)."

Children and Adolescents

- "One in five children have a diagnosable mental, emotional or behavioral disorder. And up to one in 10 may suffer from a serious emotional disturbance. Seventy percent of children, however, do not receive mental health services (SGRMH, 1999)."
- "Attention deficit hyperactivity disorder is one of the most common mental disorders in children, affecting 3 to 5 percent of school-age children (NIMH, 1999)."
- "As many as one in every 33 children and one in eight adolescents may have depression (Center for Mental Health Services (CMHS), 1998)."
- "Once a child experiences an episode of depression, he or she is at risk of having another episode within the next five years (CMHS, 1998)."

- "Teenage girls are more likely to develop depression than teenage boys (NIMH, 2000)."
- "Children and teens who have a chronic illness, endure abuse or neglect, or experience other trauma have an increased risk of depression (NIMH, 2000)."
- "Suicide is the third leading cause of death for 15- to 24-year-olds and the sixth leading cause of death for 5- to 14-year-olds. The number of attempted suicides is even higher (American Academy of Child and Adolescent Psychiatry (AACAP), 1997)."
- "Studies have confirmed the short-term efficacy and safety of treatments for depression in youth (NIMH, 2000)."
- "Twenty percent of youths in juvenile justice facilities have a serious emotional disturbance and most have a diagnosable mental disorder. Up to an additional 30 percent of youth in these facilities have substance abuse disorders or co-occurring substance abuse disorders (Office of Juvenile Justice and Delinquency Prevention (OJJDP), 2000)."

Older Adults

- "Late-life depression affects about 6 million adults, but only 10 percent ever receive treatment (NMHA, 1998)."
- "Older Americans are more likely to commit suicide than any other age group. Although they constitute only 13 percent of the U.S. population, individuals age 65 and older account for 20 percent of all suicides (NIMH, 2000)."
- "At least 10 to 20 percent of widows and widowers develop clinically significant depression within one year of their spouse's death (SGRMH, 1999)."
- "Among adults age 55 and older, 11.4 percent meet the criteria for having an anxiety disorder (SGRMH, 1999)" (National Mental Health Association (2001). Did you know? Available at: http://www.nmha.org/infoctr/didyou.cfm).

References

American Academy of Child and Adolescent Psychiatry (1998). Practice Parameters for the Assessment and Treatment of Children and Adolescents with Post-traumatic Stress Disorder. *Journal of the American Academy of Child and Adolescent Psychiatry* 37 (10 Suppl): 4S–26S.

American Psychiatric Association (1994). *Diagnostic and Statistical Manual on Mental Disorders, Fourth edition (DSM-IV).* Washington, DC: American Psychiatric Press.

American Psychiatric Association Work Group on Eating Disorders (2000). Practice Guideline for the Treatment of Patients with Eating Disorders (revision). *American Journal of Psychiatry* 157 (1 Suppl): 1–39.

Andersen, A. E. (1995). Eating Disorders in Males. In *Eating Disorders and Obesity: a Comprehensive Handbook,* ed. K. D. Brownell and C. G. Fairburn, 177–187. New York: Guilford Press.

Barkley, R. A. (1996). Attention-deficit/hyperactivity Disorder. In *Child Psychopathology,* ed. E. J. Mash and R. A. Barkley, 63–112. New York: Guilford Press.

Bourdon, K. H., J. H. Boyd, D. S. Rae, et al. (1988). Gender Differences in Phobias: Results of the ECA Community Survey. *Journal of Anxiety Disorders* 2:227–241.

Bruce, B., and W. S. Agras (1992). Binge Eating in Females: A Population-based Investigation. *International Journal of Eating Disorders* 12:365–373.

Center for Mental Health Statistics (2005). Mental Health Statistics: About the Program: Collecting National Mental Health Data. http://mentalhealth.samhsa.gov/cmhs/MentalHealthStatistics/about.asp.

Conwell, Y., and D. Brent (1995). Suicide and Aging I: Patterns of Psychiatric Diagnosis. *International Psychogeriatrics* 7:149–164.

Davidson, J. R. (2000). Trauma: The Impact of Post-traumatic Stress Disorder. *Journal of Psychopharmacology* 14 (2 Suppl 1): S5–S12.

Klerman, G. L., and M. M. Weissman (1989). Increasing Rates of Depression. *Journal of the American Medical Association* 261:2229–2235.

Kulka, R. A., W. E. Schlenger, J. A. Fairbank, et al. (1988). *Contractual Report of Findings from the National Vietnam Veterans Readjustment Study.* Research Triangle Park, NC: Research Triangle Institute.

Miniño, A. M., E. Arias, K. D. Kochanek, S. L. Murphy, and B. L. Smith (2002). Deaths: Final Data for 2000. *National Vital Statistics Reports* 50 (15). Hyattsville, MD: National Center for Health Statistics.

Murray, C. J. L., and A. D. Lopez, eds. (1996). *Summary: The Global Burden of Disease: A Comprehensive Assessment of Mortality and Disability from Diseases, Injuries, and Risk Factors in 1990 and Projected to 2020.* Cambridge, MA: Published by the Harvard School of Public Health on behalf of the World Health Organization and the World Bank, Harvard University Press. http://www.who.int/msa/mnh/ems/dalys/intro.htm.

Narrow, W. E. (1998). One-year Prevalence of Mental Disorders, Excluding Substance Use Disorders, in the United States: NIMH ECA Prospective Data. Population Estimates Based on U.S. Census Estimated Residential Population Age 18 and over on July 1, 1998. Unpublished.

Narrow, W. E., D. S. Rae, D. A. Regier. (1998). NIMH Epidemiology Note: Prevalence of Anxiety Disorders. One-year Prevalence Best Estimates Calculated from ECA and NCS Data. Population Estimates Based on U.S. Census Estimated Residential Population Age 18 to 54 on July 1, 1998. Unpublished. Available at: International Society for Mental Health Online. http://www.allaboutdepression.com/gen_25.html (accessed September 22, 2006).

Regier, D. A., W. E. Narrow, D. S. Rae, et al. (1993). The De Facto Mental and Addictive Disorders Service System. Epidemiologic Catchment Area Prospective 1-year Prevalence Rates of Disorders and Services. *Archives of General Psychiatry* 50:85–94.

Regier, D. A., D. S. Rae, W. E. Narrow, et al. (1998). Prevalence of Anxiety Disorders and their Comorbidity with Mood and Addictive Disorders. *British Journal of Psychiatry Supplement* 34:24–28.

Robins, L. N., and D. A. Regier, eds. (1991). *Psychiatric Disorders in America: The Epidemiologic Catchment Area Study.* New York: Free Press.

Shaffer, D., P. Fisher, M. K. Dulcan, et al. (1996).The NIMH Diagnostic Interview Schedule for Children Version 2.3 (DISC-2.3): Description, Acceptability, Prevalence Rates, and Rerformance in the MECA Study. Methods for the Epidemiology of Child and Adolescent Mental Disorders Study. *Journal of the American Academy of Child and Adolescent Psychiatry* 35:865–877.

Spitzer, R. L., S. Yanovski, T. Wadden, et al. (1993). Binge Eating Disorder: Its Further Validation in a Multisite Study. *International Journal of Eating Disorders* 13 (2): 137–153.

Sullivan, P. F. (1995). Mortality in Anorexia Nervosa. *American Journal of Psychiatry* 52:1073–1074.

Weissman, M. M., R. C. Bland, G. J. Canino, et al. (1999) Prevalence of Suicide Ideation and Suicide Attempts in Nine Countries. *Psychological Medicine* 29: 9–17.

Wolraich, M. L., J. N. Hannah, A. Baumgaertel, and I. D. Feurer (1998). Examination of *DSM-IV* Criteria for Attention Deficit/Hyperactivity Disorder in a County-wide Sample. *Journal of Developmental and Behavioral Pediatrics* 19:162–168.

Wonderlich, S. A., and J. E. Mitchell (1997). Eating Disorders and Comorbidity: Empirical, Conceptual, and Clinical Implications. *Psychopharmacology Bulletin* 33:381–390.

7

Documents, Reports, and Nongovernmental Organizations

This chapter identifies some of the numerous documents and reports by government agencies and nonprofit organizations that provide important information regarding mental health issues. Some of the information available in these documents is reported here. The documents include the famous Epidemiological Catchment Area research studies that report on the number of people with mental health problems in the United States. The reports include the New Freedom Mental Health Commission's latest review of the status of the U.S. mental health care system and proposals to improve it, as well as the report, *Free to Choose: Transforming Behavioral Health Care to Self-Direction,* which continues the direction of the New Freedom Mental Health Commission by encouraging self-direction by the mental health consumer of their mental health care. Some nongovernmental mental health organization Web sites are also included with examples of the kind of information that may be found on them.

Achieving the Promise: Transforming Mental Health Care in America, 2003

Publisher: New Freedom Mental Health Commission (NFMHC), Substance Abuse and Mental Health Services Administration, and National Mental Health Information Center.

Available at http://www.mentalhealthcommission.gov/reports/reports.htm.

President George W. Bush announced the New Freedom Initiative in January 2001. The goal of the new initiative was to "promote increased access to educational and employment opportunities for people with disabilities." The initiative was designed to promote access to community life and to increase the use of assistive and other technologies. The last major policy aimed at increasing participation of persons with disabilities in the communities was the Americans with Disabilities Act (ADA) in 1990 and the Supreme Court's *Olmstead v. L.C.* decision, which affirmed the right to live in community settings.

The vision statement of the initiative was as follows: "We envision a future when everyone with a mental illness will recover, a future when mental illnesses can be prevented or cured, a future when mental illnesses are detected early, and a future when everyone with a mental illness at any stage of life has access to effective treatment and supports—essentials for living, working, learning, and participating fully in the community."

As the final report states, in 2002 the president identified three barriers that prevented people with mental illness from getting the care they needed:

- "Stigma that surrounds mental illnesses,
- Unfair treatment limitations and financial requirements placed on mental health benefits in private health insurance, and
- The fragmented mental health service delivery system."

The New Freedom Mental Health Commission was established to implement the New Freedom Initiative by addressing problems in the existing mental health service delivery system. Executive Order 13263 established the duties of the commission. The president charged the commission with studying the gaps and problems in the system and making recommendations to correct and improve those problems through actions of the federal government, state governments, local governments, and public and private providers.

The commission found that there were many unmet needs and many barriers that prevented people with mental illness from getting the help they needed. They found that mental illness was very common and affected almost every American family. Mental illness affected people of all racial and ethnic backgrounds and occurred at all stages of life. Schools, workplaces, and communities were all affected by mental illness.

President Bush is quoted in the report as stating, "Americans must understand and send this message: mental disability is not a scandal—it is an illness. And like physical illness, it is treatable, especially when the treatment comes early."

As the report points out, science has steadily learned more about mental health and illnesses and how to improve mental health care. A review of scientific advances regarding mental health and illness has been released by the U.S. Department of Health and Human Services (DHHS) in *Mental Health: A Report of the Surgeon General.* Unfortunately, many Americans still do not benefit from this increased knowledge and the more effective treatments. "Far too often, treatments and services that are based on rigorous clinical research languish for years rather than being used effectively at the earliest opportunity. For instance, according to the Institute of Medicine report, *Crossing the Quality Chasm: A New Health System for the 21st Century,* the lag between discovering effective forms of treatment and incorporating them into routine patient care is unnecessarily long, lasting about 15 to 20 years."

Although mental illnesses rank first among illnesses that cause disability in the United States, Canada, and Western Europe, this fact is poorly recognized in terms of its public health burden. "In addition, one of the most distressing and preventable consequences of undiagnosed, untreated, or under-treated mental illnesses is suicide." The World Health Organization (WHO) reports that suicide worldwide causes more deaths every year than war or homicide.

Mental illnesses also come with a very high financial cost. "In the U.S., the annual economic, indirect cost of mental illnesses is estimated to be $79 billion. Most of that amount—approximately $63 billion—reflects the loss of productivity as a result of illnesses. But indirect costs also include almost $12 billion in mortality costs (lost productivity resulting from premature death) and almost $4 billion in productivity losses for incarcerated individuals and for the time of those who provide family care. In 1997, the latest year comparable data are available, the United States spent more than $1 trillion on health care, including almost $71 billion on treating mental illnesses. Mental health expenditures are predominantly publicly funded at 57%, compared to 46% of overall health care expenditures. Between 1987 and 1997, mental health spending did not keep pace with general health care because of declines in private health spending under managed care and cutbacks in hospital expenditures. "

In its *Interim Report to the President,* the commission declared, "the mental health delivery system is fragmented and in disarray . . . lead[ing] to unnecessary and costly disability, homelessness, school failure and incarceration." The report detailed barriers and unmet needs including:

- "Fragmentation and gaps in care for children,
- Fragmentation and gaps in care for adults with serious mental illnesses,
- High unemployment and disability for people with serious mental illnesses,
- Lack of care for older adults with mental illnesses, and
- Lack of national priority for mental health and suicide prevention."

The *Interim Report* concludes that the mental health care system does not meet the most important goal of people with mental illness—recovery. Research has led to many new and improved treatments, but there is a failure to transfer that research into community treatment settings. Access to quality care in many communities is lacking. Improved treatment and supports in the communities would result in more individuals, even with serious mental illnesses, being able to recover. The commission did not blame the caregivers, but placed the blame on how the mental health care system had developed. The commission stressed the need to replace unnecessary institutional care with community care and stressed the need to integrate fragmented programs.

The commission, during its year-long study, reviewed the scientific literature and the comments of more than 2,300 consumers, family members, providers, administrators, researchers, government officials, and others on how mental health care is delivered. The commission concluded that traditional reform measures were not sufficient to meet the expectations of consumers and families. The commission recommended fundamentally changing how mental health care is delivered in the United States.

The report calls for new service delivery patterns and incentives to bring about easy and continuous access for everyone. "When a serious mental illness or a serious emotional disturbance is first diagnosed, the health care provider—in full partnership with consumers and families—will develop an individualized plan of care for managing the illness. This partnership of personalized care means basically choosing *who, what,* and *how* appropriate health care will be provided:

- Choosing which mental health care professionals are on the team,
- Sharing in decision making, and
- Having the option to agree or disagree with the treatment plan."

The report calls for recovery to be the recognized outcome of mental health services. Successfully transforming the system is seen as requiring adherence to two principles:

- "First, services and treatments must be consumer and family centered, geared to give consumers real and meaningful choices about treatment options and providers—not oriented to the requirements of bureaucracies.
- Second, care must focus on increasing consumers' ability to successfully cope with life's challenges, on facilitating recovery, and on building resilience, not just on managing symptoms."

The report calls for reimbursement to be based on treatments and services with proven effectiveness and consumer preference and for investment in infrastructure to support new technologies. Table 7.1 shows the six goals established by the commission for transforming the system.

TABLE 7.1

Goals and Recommendations for Implementation in a Transformed Mental Health System

Goal 1: Americans understand that mental health is esential to overall health.	
Recommendations	1.1 Advance and implement a national campaign to reduce the stigma of seeking care and a national strategy for suicide prevention.
	1.2 Address mental health with the same urgency as physical health.
Goal 2: Mental health care is consumer and family driven.	
Recommendations	2.1 Develop an individualized plan of care for every adult with a serious mental illness and child with a serious emotional disturbance.
	2.2 Involve consumers and families fully in orienting the mental health system toward recovery.
	2.3 Align relevant Federal programs to improve access and accountability to mental health services.
	2.4 Create a comprehensive State Mental Health Plan.
	2.5 Protect and enhance the rights of people with mental illnesses.
Goal 3: Disparities in mental health services are eliminated.	
Recommendations	3.1 Improve access to quality care that is culturally competent.
	3.2 Improve access to quality care in rural and geographically remote areas.
Goal 4: Early mental health screening, assessment, and referral to services are common practice.	
Recommendations	4.1 Promote the mental health of young children.
	4.2 Improve and expand school mental health programs.
	4.3 Screen for co-occurring mental and substance use disorders and link with integrated treatment strategies.
	4.4 Screen for mental disorders in primary health care, across the life span, and connect to treatment and supports.
Goal 5: Excellent mental health care is delivered and research is accelerated.	
Recommendations	5.1 Accelerate research to promote recovery and resilience, and ultimately to cure and prevent mental illnesses.
	5.2 Advance evidence-based practices using dissimenation and demonstration projects and create a public-private partnership to guide their implementation.
	5.3 Improve and expand the workforce providing evidence-based mental health services and supports.
	5.4 Develop the knowledge base in four understudied areas: mental health disparities, long-term effects of medications, trauma, and acute care.
Goal 6: Technology is used to access mental health care and information.	
Recommendations	6.1 Use health technology and telehealth to improve access and coordination of mental health care, especially for Americans in remote areas or in underserved populations.
	6.2 Develop and implement integrated electronic health record and personal health information systems.

Source: President's New Freedom Mental Health Commission, 2003.

Epidemiological Studies: Epidemiological Catchment Area (ECA) Study 1980–1985, National Comorbidity Study 1990–1992, and NCS-Replication Released 2005

Available at: http://www.hcp.med.harvard.edu/ncs/; http://www.hcp.med.harvard.edu/ncs/replication.php.

In 1977 the President's Commission on Mental Health called for research to determine the extent of mental health problems in the United States. In response to that report, in the early 1980s, the National Institute of Mental Health in the Department of Health and Human Services conducted a series of studies, the Epidemiologic Catchment Area (ECA) studies, on the prevalence of mental illness in the United States. Data was collected on the incidence and prevalence of mental disorders in the United States and on the use of services and the need for additional services by the mentally ill. The National Institute for Mental Health collaborated with research teams from the University of California at Los Angeles, Washington University, Duke University, Johns Hopkins University, and Yale University. The research was conducted in five community mental health center catchment areas: Los Angeles, California; St. Louis, Missouri; Durham, North Carolina; Baltimore, Maryland; and New Haven, Connecticut.

The same questions and sample characteristics were used for each site. At each site, 3,000 community residents and 500 residents of institutions were sampled, for a total of 20,861 respondents. Three interviews were conducted over a one-year time period, two personal interviews and, for the household sample, a brief telephone interview midway between the two personal interviews. The interviews used the National Institute of Mental Health's Diagnostic Interview Schedule III and the diagnoses were categorized using the *Diagnostic and Statistical Manual of Mental Disorders,* 3rd edition. Diagnoses included manic episode, bipolar disorder, atypical bipolar disorder, single-episode major depression, dysthymia, recurrent major depression, schizophrenia, schizophreniform, phobia, obsessive compulsive disorder, panic, somatization, antisocial personality, anorexia nervosa, alcohol abuse or dependence, and drug abuse or dependence. The ECA study was limited, because the data could not be generalized from the five study sites to the general population.

The second study, the National Comorbidity Survey (NCS), analyzed a national representative sample in 1990–1992 using the World Health Organization's Composite International Diagnostic Interview (CIDI), which was based on the Diagnostic and Statistical Manual III-R (DSM-III-R). The third study, the NCS-Replication, examined trends in mental health prevalence, impairment, and service and was released in 2005. That study consisted of a representative sample of the U.S. population using many of the questions from the original study but using assessment criteria based on Diagnostic and Statistical Manual IV (DSM-IV). The World Health Organization's World Mental Health Survey Initiative was used as a guide in collecting the data. The WHO data gathers cross-national data from twenty-eight countries from all WHO regions. Information is collected on mental, behavior, and substance use disorders to identify barriers to service, evaluate intervention targets, and estimate the world mental health burden. These reports provide perhaps the best data in the United States regarding the numbers and types of mental disorders in the United States.

Free to Choose: Transforming Behavioral Health Care to Self-Direction: Report of the 2004 Consumer Direction Initiative Summit

Publisher: Substance Abuse and Mental Health Services Administration 2005. DHHS Publication No. SMA-05-3982, printed 2005.

Available at: http://mentalhealth.samhsa.gov/publications/allpubs/SMA05-3982/default.asp.

The report found that self-directed health care is becoming a significant element for change in the American health care system. This trend is being driven by increases in the cost of health services and health insurance and the inconsistent quality of the health care provided. President Bush's New Freedom Commission on Mental Health supports community-oriented services in behavioral health and development of consumer-driven models of care.

The Substance Abuse and Mental Health Services Administration (SAMHSA) in the Department of Health and Human Services has supported consumer-operated services, peer recovery programs, and consumer advocacy. In 2003 SAMHSA recommended, through a self-determination planning meeting, development of a consumer-directed health care initiative for persons with mental illness and/or substance abuse disorders. This recommendation was followed up in 2004 with a Consumer Direction Initiative Summit meeting in Washington, D.C. The seventy-nine participants of the summit were consumers of mental health and addiction services, family members, providers, policy makers, and federal and state representatives. Workgroups identified needs of consumers and potential barriers to self-directed behavioral health care. A vision was developed, and recommendations were made for changing the delivery system. The recommendations were presented to the Center for Mental Health Services (CMHS), the Center for Substance Abuse Treatment (CSAT), and the Center for Substance Abuse Prevention (CSAP). The working papers are available on the Web at www.mentalhealth.samhsa.gov/consumersurvivor under "Featured Publications." The following paragraphs provide a summary of some of the topics and themes from the papers.

In the report, the term *self-directed care* refers to a system that is "intended to allow informed consumers to assess their own needs . . . determine how and by whom these needs should be met, and monitor the quality of services they receive" (Dougherty 2003). It refers to a system "in which funds that would ordinarily be paid to service provider agencies are transferred to consumers, using various formulas to account for direct, administrative, and other costs" (Cook, Terrell, and Jonikas 2004).

Four essential elements of self-directed care are identified by the Federal Centers for Medicare and Medicaid Services (CMS) (Cook, Terrell, and Jonikas 2004):

- "Person-centered planning, which constitutes a comprehensive strategy for putting necessary services and supports in place to help people achieve their goals;
- Individual budgeting, which enables people needing assistance to have some control over how the funds used for their care are to be spent;

- Financial management services, which encompass such activities as tracking and monitoring budgets, performing payroll services, and handling billing and documentation; and
- Supports brokerage, which includes both education and operational assistance, and is intended to help participants design and manage their self-directed care plans."

Values for a recovery-based mental health system were defined as the following:

- "Self-determination;
- Empowering relationships based on trust, understanding, and respect;
- Meaningful roles in society; and
- Elimination of stigma and discrimination."

The report encouraged changing the mental health system to one based on recovery through the building of alliances of consumers and allies.

Change is supported by the President's New Freedom Commission on Mental Health, whose report, discussed above, called for a transformation of the mental health system to one based on the principles of recovery, as stated in its vision: "We envision a future when everyone with a mental illness will recover." The Surgeon General's report on mental health (U.S. Department of Health and Human Services 2000) and the National Council on Disabilities report, *From Privileges to Rights* (National Council on Disabilities 2000), also emphasize the importance of recovery from mental illness in the policies and services of the system.

Global Burden of Disease

Publisher: Harvard School of Public Health on behalf of World Health Organization (WHO), Geneva, Switzerland, and World Bank, Washington, DC, 2002.

Available at: http://www.who.int/en/; http://www.who.int/healthinfo/bodproject/en/.

"The term 'burden of disease' can refer to the overall impact of diseases and injuries at the individual level, at the societal level, or to the economic costs of diseases. Specifically, the 'global burden of disease' (GBD) refers to a WHO and World Bank study published in the World Development Report 1993 that measured the total loss of health resulting from diseases and injuries" (WHO 2002). The study showed that infectious disease accounts for 43 percent of the global burden. The study was updated in 1996 and 2000. These three studies have provided the most consistent and comprehensive estimates on morbidity and mortality worldwide by region, age, and sex.

"The GBD study also introduced a new metric disability-adjusted life year (DALY) to quantify the burden of disease. DALYs are used to help measure the burden of disease and the effectiveness of health interventions. The DALY is a health gap measure, which combines information on the impact of premature death, and of disability and other non-fatal health outcomes. Also statistically innovative are years of life lost (YLLs), an indicator showing loss from premature mortality, based on early death judged against the average life expectancy in the population of a developed country. Thus the burden of disease is a measurement of the gap between current health status and an ideal situation where everyone lives into old age, free of disease and disability. In the World Health Report 2000, WHO introduced the disability-adjusted life expectancy (DALE), a summary measure of the level of health attained by populations. The name of this measure was changed to health-adjusted life expectancy (HALE) in the World Health Report 2002, to better reflect the inclusion of all states of health in the calculation. The HALE is based on life expectancy at birth but includes an adjustment for time spent in poor health. It is most easily understood as the equivalent number of years in full health that a newborn can expect to live, based on current rates of ill health and mortality" (WHO 2002).

Burden of disease statistics were used by the Commission on Macroeconomics and Health to show that ill health among the poor is the result of a relatively small number of identifiable conditions, which include childhood infectious diseases, HIV/AIDS, malaria, maternal and perinatal conditions, micronutrient deficiencies, tobacco-related illnesses, and tuberculosis.

Priority settings may be assisted by the use of DALYs and HALE. Priority setting allows policy makers to rank health problems and allocate funds to address those problems. They are able to take into consideration both the disease burden and the cost-effectiveness of various approaches to address the disease burden. They may also consider other social goals, such as reducing inequality in health. "For example, in some countries, five out of the ten leading causes of DALYs are related to reproductive ill-health, suggesting that this should be a priority in these countries, assuming the existence of cost-effective interventions. Some are concerned that this approach leads to a focus on a single disease or group of diseases, with the result that complementary efforts required to strengthen public health systems are neglected. However, GBD analysts have never suggested that burden alone should be used to set priorities" (WHO 2002).

The global burden of disease study included mental illness, and perhaps the most powerful finding of the study was that the impact of mental illness on overall health and productivity had been profoundly underrecognized. All cardiovascular conditions accounted for 18.6 percent of total DALYs while mental disorders accounted for 15.4 percent of DALYs, slightly more than the burden associated with all forms of cancer (see Table 7.2).

TABLE 7.2
Disease Burden by Selected Illness Categories in Established Market Economies, 1990

	Precent of total DALYs*
All cardiovascular conditions	18.6
All mental illness**	15.4
All malignant disease (cancer)	15.0
All respiratory conditions	4.8
All alcohol use	4.7
All infectious and parasitic disease	2.8
All drug use	1.5

*Disability-adjusted life year (DALY) is a measure that expresses years of life lost to premature death and years lived with a disability of specified severity and duration (Murray and Lopez, WHO, 1996).

**Disease burden associated with "mental illness" includes suicide (Murray and Lopez, WHO, 1996).

Mental Health: A Report of the Surgeon General—Executive Summary

Author: Surgeon General, U.S. Public Health Service, U.S. Department of Health and Human Services.

Publisher: Rockville, MD: U.S. Department of Health and Human Services, Substance Abuse and Mental Health Services Administration, Center for Mental Health Services, National Institutes of Health, National Institute of Mental Health, 1999.

Available at http://www.surgeongeneral.gov/library/mentalhealth/home.html.

According to the executive summary of the Surgeon General's report, the report begins by defining mental health and mental illness:

> "Mental health—the successful performance of mental function, resulting in productive activities, fulfilling relationships with other people, and the ability to adapt to change and to cope with adversity; from early childhood until late life, mental health is the springboard of thinking and communication skills, learning, emotional growth, resilience, and self-esteem.
>
> Mental illness—the term that refers collectively to all mental disorders. Mental disorders are health conditions that are characterized by alterations in thinking, mood, or behavior (or some combination thereof) associated with distress and/or impaired functioning."

This report was the first Surgeon General's report ever issued on the topic of mental health and mental illness. The report was based on current science, and the executive summary summed up its basic points:

- "One is that mental health is fundamental to health. The qualities of mental health are essential to leading a healthy life. Americans assign high priority to preventing disease and promoting personal well-being and public health; so too must we assign priority to the task of promoting mental health and preventing mental disorders. Nonetheless, mental disorders occur and, thus, treatment and mental health services are critical to the Nation's health. These emphases, combined with research to increase the knowledge needed to treat and prevent mental and behavioral disorders, constitute a broad public health approach to an urgent health concern.

- A second message of the report is that mental disorders are real health conditions that have an immense impact on individuals and families throughout this Nation and the world. Appreciation of the clinically and economically devastating nature of mental disorders is part of a quiet scientific revolution that not only has documented the extent of the problem, but also in recent years has generated many real solutions. The decision to publish the report at this time was based, in part, on the tremendous growth of the science base that is enriching our understanding of the complexity of the brain and behavior. This understanding increasingly supports mental health practices."

The report was based on an extensive review of the scientific literature covering more than 3,000 research articles and other materials and on consultations with mental health care providers and consumers. The report found a strong consensus among Americans that society could no longer afford to view mental health as separate from and unequal to general health. The Surgeon General's report found that mental health should be part of the mainstream of health.

The review of research supported two main findings:

- "The efficacy of mental health treatments is well documented, and
- A range of treatments exists for most mental disorders."

Based on the findings a single recommendation was made: "Seek help if you have a mental health problem or think you have symptoms of a mental disorder."

The report discusses the various treatment options available for the consumer to choose from, but states that still nearly half of all Americans with a severe mental illness do not seek treatment. The report notes that among the barriers to seeking help, stigma is still the greatest barrier.

> "Stigma erodes confidence that mental disorders are valid, treatable health conditions. It leads people to

avoid socializing, employing or working with, or renting to or living near persons who have a mental disorder, especially a severe disorder like schizophrenia. Stigma deters the public from wanting to pay for care and, thus, reduces consumers' access to resources and opportunities for treatment and social services. A consequent inability or failure to obtain treatment reinforces destructive patterns of low self-esteem, isolation, and hopelessness. Stigma tragically deprives people of their dignity and interferes with their full participation in society. Increasingly effective treatments for mental disorders promise to be the most effective antidote to stigma. Effective interventions help people to understand that mental disorders are not character flaws but are legitimate illnesses that respond to specific treatments, just as other health conditions respond to medical interventions."

The report has a series of key themes:

- "The importance of information, policies, and actions that will reduce and eventually eliminate the cruel and unfair stigma attached to mental illness.
- The importance of a solid research base for every mental health and mental illness intervention.
- Public Health Perspective—In the United States, mental health programs, like general health programs, are rooted in a population-based public health model. Broader in focus than medical models that concentrate on diagnosis and treatment, public health attends, in addition, to the health of a population in its entirety. A public health approach encompasses a focus on epidemiologic surveillance, health promotion, disease prevention, and access to services. Although much more is known through research about mental illness than about mental health, the report attaches high importance to public health practices that seek to identify risk factors for mental health problems; to mount preventive interventions that may block the emergence of severe illnesses; and to actively promote good mental health.
- Mental Disorders Are Disabling—The World Health Organization, in collaboration with the World Bank and Harvard University, mounted an ambitious research

effort in the mid-1990s to determine the 'burden of disability' associated with the whole range of diseases and health conditions suffered by peoples throughout the world. . . . Today, in established market economies such as the United States, mental illness is the second leading cause of disability and premature mortality.

- Mental Health and Mental Illness: Points on a Continuum—'mental health' and 'mental illness' may be thought of as points on a continuum. *Mental health* refers to the successful performance of mental function, resulting in productive activities, fulfilling relationships with other people, and the ability to adapt to change and to cope with adversity. Mental health is indispensable to personal well-being, family and interpersonal relationships, and contribution to community or society. It is easy to overlook the value of mental health until problems surface. Yet from early childhood until death, mental health is the springboard of thinking and communication skills, learning, emotional growth, resilience, and self-esteem. These are the ingredients of each individual's successful contribution to community and society. Americans are inundated with messages about *success*—in school, in a profession, in parenting, in relationships—without appreciating that successful performance rests on a foundation of mental health.
- *Mental illness* refers collectively to all diagnosable mental disorders. Mental disorders are health conditions that are characterized by alterations in thinking, mood, or behavior (or some combination thereof) associated with distress and/or impaired functioning.
- Mind and Body Are Inseparable—As it examines mental health and illness in the United States, the report confronts a profound obstacle to public understanding, one that stems from an artificial, centuries-old separation of mind and body."

Even today, everyday language encourages a misperception that mental health or mental illness is unrelated to physical health or physical illness. In fact, the two are inseparable. In keeping with modern scientific thinking, this report uses mind to refer to all mental functions related to thinking, mood, and purposive behavior. The mind is generally seen as deriving from activities within the brain. Research reviewed for this report makes it clear that mental functions are carried out by a particular organ, the brain. "Indeed, new and emerging technologies are making it increasingly possible for researchers to demonstrate the extent to which mental disorders and their treatment—both with medication and with psychotherapy—are reflected in physical changes in the brain."

The report is divided into several chapters. Chapter 1, "Introduction and Themes," summarizes five overarching themes: mental health and mental illness as a public health approach, mental disorders as disabling, mental health and mental illness as points on a continuum, mind and body as inseparable, and the roots of stigma.

Chapter 2, "The Fundamentals of Mental Health and Mental Illness," addresses the defining trends in the mental health field of the past twenty-five years, including the following:

- "The extraordinary pace and productivity of scientific research on the brain and behavior;
- The introduction of a range of effective treatments for most mental disorders;
- A dramatic transformation of our society's approaches to the organization and financing of mental health care; and
- The emergence of powerful consumer and family movements."

Conclusions found in Chapter 2 include:

- "The multifaceted complexity of the brain is fully consistent with the fact that it supports all behavior and mental life. Proceeding from an acknowledgment that all psychological experiences are recorded ultimately in the brain and that all psychological phenomena reflect biological processes, the modern neuroscience of mental health offers an enriched understanding of the inseparability of human experience, brain, and mind.
- Mental functions, which are disturbed in mental disorders, are mediated by the brain. In the process of transforming human experience into physical events, the

brain undergoes changes in its cellular structure and function.

- Few lesions or physiologic abnormalities define the mental disorders, and for the most part their causes remain unknown. Mental disorders, instead, are defined by signs, symptoms, and functional impairments.
- Diagnoses of mental disorders made using specific criteria are as reliable as those for general medical disorders.
- About one in five Americans experiences a mental disorder in the course of a year. Approximately 15 percent of all adults who have a mental disorder in one year also experience a co-occurring substance (alcohol or other drug) use disorder, which complicates treatment.
- A range of treatments of well-documented efficacy exists for most mental disorders. Two broad types of intervention include psychosocial treatments—for example, psychotherapy or counseling—and psychopharmacologic treatments; these often are most effective when combined.
- In the mental health field, progress in developing preventive interventions has been slow because, for most major mental disorders, there is insufficient understanding about etiology (or causes of illness) and/or there is an inability to alter the *known* etiology of a particular disorder. Still, some successful strategies have emerged in the absence of a full understanding of etiology.
- About 10 percent of the U.S. adult population use mental health services in the health sector in any year, with another 5 percent seeking such services from social service agencies, schools, or religious or self-help groups. Yet critical gaps exist between those who need service and those who receive service.
- Gaps also exist between optimally effective treatment and what many individuals receive in actual practice settings.
- Mental illness and less severe mental health problems must be understood in a social and cultural context, and mental health services must be designed and delivered in a manner that is sensitive to the perspectives and needs of racial and ethnic minorities.

- The consumer movement has increased the involvement of individuals with mental disorders and their families in mutual support services, consumer-run services, and advocacy. They are powerful agents for changes in service programs and policy.
- The notion of recovery reflects renewed optimism about the outcomes of mental illness, including that achieved through an individual's own self-care efforts, and the opportunities open to persons with mental illness to participate to the full extent of their interests in the community of their choice."

Chapter 3, "Children and Mental Health," Chapter 4, "Adults and Mental Health," and Chapter 5, "Older Adults and Mental Health," look at mental health and mental illness across the lifespan. The periods covered are childhood and adolescence, adulthood, and later adult life beginning somewhere between ages fifty-five and sixty-five.

Frequency of occurrence and economic, societal, and clinical burden were taken into consideration in selecting disorders for discussion. Where data was available, gender and culture were considered. The chapters also consider diagnosis, course of the disease, and treatment as well as the role of consumers and families.

Chapter 6, "Organizing and Financing Mental Health Services," addresses organization and financing of mental health services. The report indicates that research on identification, treatment, and in some cases prevention is exceeding the ability of the existing service system to deliver services: "Approximately 10 percent of children and adults receive mental health services from mental health specialists or general medical providers in a given year. Approximately one in six adults, and one in five children, obtain mental health services either from health care providers, the clergy, social service agencies, or schools in a given year."

Concerns are raised about possible discrimination in mental health financing. Managed care has become a way to try to contain costs. "Intensive research currently is addressing both positive and adverse effects of managed care on access and quality, generating information that will guard against untoward consequences of aggressive cost-containment policies. Inequities in insurance coverage for mental health and general medical care—the product of decades of stigma and discrimination—have prompted efforts to correct them through legislation designed to produce financing changes and create parity. Parity calls for equality between mental health and other health coverage."

Chapter 7, "Actions for Mental Health in the New Millennium," ends the report. It concludes that "*a range of treatments of documented efficacy exists for most mental disorders*. Moreover, a person may choose a particular approach to suit his or her needs and preferences. Based on this finding, the report's principal recommendation to the American people is to *seek help if you have a mental health problem or think you have symptoms of a mental disorder.*" The chapter also notes that stigma is still a major barrier for people seeking help, and there are still substantial gaps in the services available.

National Alliance for the Mentally Ill

Available at http://www.nami.org/.

The following is taken from the Web site of the National Alliance for the Mentally Ill (NAMI). NAMI's mission statement says: "NAMI is dedicated to the eradication of mental illnesses and to the improvement of the quality of life of all whose lives are affected by these diseases."

NAMI is a nonprofit advocacy organization of consumers, families, and friends of people with severe mental illnesses. NAMI membership covers a broad range of mental disorders including schizophrenia, schizoaffective disorder, bipolar disorder, major depressive disorder, panic and other anxiety disorders, obsessive-compulsive disorder, autism and pervasive developmental disorders, attention deficit hyperactivity disorder, and other severe and persistent mental illnesses that affect the brain.

The National Alliance for the Mentally Ill was founded in 1979. NAMI seeks to improve treatment and services for the more than 15 million Americans with severe mental illness and their families. NAMI has more than one thousand local affiliates and 50 state organizations.

The national office, local affiliates, and state organizations identify and work on issues most important to their community and state. The national office provides leadership, advocacy, and education at the national level, and provides direction to the entire organization. The state organizations and local affiliates work on issues important to the state and local levels.

Nonprofit organizations such as NAMI are another source of mental health information and data. NAMI makes available the following information on their Web site:

- "Mental illnesses include such disorders as schizophrenia, schizoaffective disorder, bipolar disorder, major depressive disorder, obsessive-compulsive disorder, panic and other severe anxiety disorders, autism and pervasive developmental disorders, attention deficit/hyperactivity disorder, borderline personality disorder, and other severe and persistent mental illnesses that affect the brain.
- These disorders can profoundly disrupt a person's thinking, feeling, moods, ability to relate to others and capacity for coping with the demands of life. Mental illnesses can affect persons of any age, race, religion, or income. Mental illnesses are not the result of personal weakness, lack of character, or poor upbringing.
- Mental illnesses are treatable. Most people with serious mental illness need medication to help control symptoms, but also rely on supportive counseling, self-help groups, assistance with housing, vocational rehabilitation, income assistance and other community services in order to achieve their highest level of recovery."

Here are some important facts from NAMI about mental illness and recovery:

- "Mental illnesses are biologically based brain disorders. They cannot be overcome through 'will power' and are not related to a person's 'character' or intelligence.
- Mental disorders fall along a continuum of severity. The most serious and disabling conditions affect five to ten million adults (2.6–5.4%) and three to five million children ages five to seventeen (5–9%) in the United States.
- Mental disorders are the leading cause of disability (lost years of productive life) in North America, Europe and, increasingly, in the world. By 2020, Major Depressive illness will be the leading cause of disability in the world for women and children.

- Mental illnesses strike individuals in the prime of their lives, often during adolescence and young adulthood. All ages are susceptible, but the young and the old are especially vulnerable.
- Without treatment the consequences of mental illness for the individual and society are staggering: unnecessary disability, unemployment, substance abuse, homelessness, inappropriate incarceration, suicide and wasted lives. The economic cost of untreated mental illness is more than 100 billion dollars each year in the United States.
- The best treatments for serious mental illnesses today are highly effective; between 70 and 90 percent of individuals have significant reduction of symptoms and improved quality of life with a combination of pharmacological and psychosocial treatments and supports.
- Early identification and treatment is of vital importance. By getting people the treatment they need early, recovery is accelerated and the brain is protected from further harm related to the course of illness.
- Stigma erodes confidence that mental disorders are real, treatable health conditions."

National Mental Health Association

Available at http://www.who.int/whr/2001/overview/en/index.html.

Founded in 1909 by Clifford W. Beers, the National Mental Health Association is the oldest and largest nonprofit organization covering a wide range of issues regarding mental health and mental illness. The association has 340 affiliates throughout the country and is involved in service, research, advocacy, and education.

The National Mental Health Association (NMHA) is one of the oldest and largest mental health organizations in the United States and a source for information on mental health. NMHA sums up its vision thus: "The National Mental Health Association envisions a just, humane and healthy society in which all people are accorded respect, dignity, and the opportunity to achieve their full potential free from stigma and prejudice. Justice demands that every person, regardless of disability and other characteristics such as race, ethnicity, gender, age, economic status or sexual orientation, has the right and responsibilities of full participation in society."

The goals of NMHA are as follows:

- *Advocacy:* Through powerful and effective advocacy, NMHA and its affiliates will craft and support public policies that promote mental health, consumer empowerment, and ensure an integrated, comprehensive and accessible system of care for children, adults, and seniors.
- *Public Education:* NMHA and its affiliates will reduce stigma and improve public understanding, attitudes, and actions regarding mental health and mental illnesses.
- *Research:* NMHA and its affiliates will work to ensure the development of a broad-based national mental health research agenda, which includes basic research, services research, and prevention research.
- *Services:* NMHA will facilitate and support Mental Health Association affiliate efforts to provide high quality, culturally competent mental health services and support to children, adults, and seniors according to local needs and affiliate capacity.
- *Organization:* NMHA and its affiliates will enhance the resources and infrastructure necessary to support and advance the NMHA mission.

National Strategy for Suicide Prevention: Goals and Objectives for Action

Author: Surgeon General, Public Health Service, Department of Health and Human Services, 2001.

Available at http://www.ncbi.nlm.nih.gov/books/bv.fcgi?rid=hstat5.chapter.4214.

The following information is taken from the *National Strategy for Suicide Prevention* (*National Strategy* or NSSP). NSSP is a strategy to reduce the rate of suicide in the United States. It is designed to require a variety of organizations and individuals to become involved in suicide prevention at all levels of government—federal, state, tribal, and community. NSSP is the first attempt in the United States to prevent suicide through a coordinated approach by both the public and private sectors. The strategy establishes goals and objectives to reduce the rate of suicide.

The plan is based on research about suicidal behavior and suicide prevention and uses a public health approach.

Aims of the *National Strategy*

- "Prevent premature deaths due to suicide across the life span,
- Reduce the rates of other suicidal behaviors,
- Reduce the harmful after-effects associated with suicidal behaviors and the traumatic impact of suicide on family and friends, and
- Promote opportunities and settings to enhance resiliency, resourcefulness, respect, and interconnectedness for individuals, families, and communities."

A Plan for Suicide Prevention: Goals, Objectives, and Activities

The *National Strategy* has eleven goals and sixty-eight objectives. The eleven goals are:

> "Goal 1: Promote awareness that suicide is a public health problem that is preventable.
>
> Goal 2: Develop broad-based support for suicide prevention.
>
> Goal 3: Develop and implement strategies to reduce the stigma associated with being a consumer of mental health, substance abuse and suicide prevention services.
>
> Goal 4: Develop and implement suicide prevention programs.
>
> Goal 5: Promote efforts to reduce access to lethal means and methods of self-harm.

Goal 6: Implement training for recognition of at-risk behavior and delivery of effective treatment.

Goal 7: Develop and promote effective clinical and professional practices.

Goal 8: Improve access to and community linkages with mental health and substance abuse services.

Goal 9: Improve reporting and portrayals of suicidal behavior, mental illness, and substance abuse in the entertainment and news media.

Goal 10: Promote and support research on suicide and suicide prevention.

Goal 11: Improve and expand surveillance systems."

The *National Strategy* takes a public health approach with prevention efforts, focusing on identifying patterns of suicide and suicidal behavior throughout a population. Surveillance is conducted to determine the rates of suicide and suicidal behavior. Information is collected on "the characteristics of individuals who die by suicide, the circumstances surrounding these incidents, possible precipitating events, and the adequacy of social support and health services." Data may also be collected on the costs of injuries resulting from suicidal behavior, but accurate data on attempted suicides is much more difficult to obtain. Surveillance helps the community to define the extent of the problem and changes in suicide rates over time by geographic region, age groups, race, ethnicity, and gender.

Not all deaths that are suicides are reported as suicides. Sometimes they are reported as accidents or homicides. "Suicide rates have changed over time, especially among certain subgroups. For example, from 1980 to 1996, the rate of suicide among children aged 10–14 increased by 100 percent, and among African-American males aged 15–19, the rate increased by 105 percent" (Peters et al. 1998).

There is no national database on attempted suicide. The Centers for Disease Control biennially conducts a survey on attempted suicide by youth. The Youth Risk Behavior Survey shows that a large number of youth in grades 9 through 12 consider or attempt suicide.

Suicide is believed to be very costly to the nation, but due to incomplete data it is difficult to accurately assess the cost of suicide. "One economic analysis, however, estimated the total economic burden of suicide in the United States in 1995 to be $111.3 billion; this includes medical expenses of $3.7 billion, work-related losses of $27.4 billion, and quality of life costs of $80.2 billion."

"Risk factors may be thought of as leading to or being associated with suicide; that is, people 'possessing' the risk factor are at greater potential for suicidal behavior. Protective factors, on the other hand, reduce the likelihood of suicide. They enhance resilience and may serve to counterbalance risk factors. Risk and protective factors may be biopsychosocial, environmental, or sociocultural in nature. Although this division is somewhat arbitrary, it provides the opportunity to consider these factors from different perspectives." Suicide prevention programs can benefit from knowing the risk and protective factors for various groups. Following are some of the protective factors for suicide:

- "Effective clinical care for mental, physical, and substance use disorders,
- Easy access to a variety of clinical interventions and support for help-seeking,
- Restricted access to highly lethal means of suicide,
- Strong connections to family and community support,
- Support through ongoing medical and mental health care relationships,
- Skills in problem solving, conflict resolution, and nonviolent handling of disputes, and
- Cultural and religious beliefs that discourage suicide and support self-preservation."

There are also risk factors for suicide including the following:

"*Biopsychosocial Risk Factors*
- Mental disorders, particularly mood disorders, schizophrenia, anxiety disorders and certain personality disorders,
- Alcohol and other substance use disorders,
- Hopelessness,
- Impulsive and/or aggressive tendencies,
- History of trauma or abuse,
- Some major physical illnesses,

- Previous suicide attempt, and
- Family history of suicide.

Environmental Risk Factors

- Job or financial loss,
- Relational or social loss,
- Easy access to lethal means, and
- Local clusters of suicide that have a contagious influence.

Sociocultural Risk Factors

- Lack of social support and sense of isolation,
- Stigma associated with help-seeking behavior,
- Barriers to accessing health care, especially mental health and substance abuse treatment,
- Certain cultural and religious beliefs (for instance, the belief that suicide is a noble resolution of a personal dilemma), and
- Exposure to, including through the media, and influence of others who have died by suicide."

Suicide prevention interventions are designed either to increase protective factors or to reduce risk, and in some cases to do both. An intervention can be developed to impact the physical environment, a psychological state, or a culture. Suicide prevention can be universal, aimed at affecting the whole population, or selective, aimed at affecting a subgroup. It may also be "indicated" as for a person with a risk factor.

The section entitled "Matrix of Interventions for Suicide Prevention" provides a matrix that shows a mix of suicide interventions. State and local organizations can use a variety of those interventions to establish suicide prevention programs. Each locality can select interventions that best address its local needs.

Comprehensive suicide prevention programs are believed to be more likely to have an impact on the suicide rate than programs that only address one protection or risk factor. Comprehensive programs are designed to involve a wide range of community resources, including public health, health and mental health care, education, social services, law enforcement, civic groups, business, and faith communities. It is also important to evaluate interventions because most interventions, even those that have been widely used, have not been evaluated. The United

States is one of a small number of countries that has a national strategy to address suicide. The U.S. approach is based on the 1996 publication, *Prevention of Suicide: Guidelines for the Formulation and Implementation of National Strategies,* published by the World Health Organization of the United Nations.

Finland, which experiences a high suicide rate, was the first country to establish a national suicide prevention strategy in 1986. The U.S. strategy has benefited from the experience of Finland, Norway, Sweden, the United Kingdom, the Netherlands, Estonia, France, Australia, and New Zealand. Some of the common elements in national strategies are education, changes in media portrayal of suicide and mental illness, increased detection and treatment of depression and substance abuse disorders, reduction of stigma, improved clinical practices, and reduction of access to lethal means of suicide.

Effective suicide prevention programs:

- "Clearly identify the population that will benefit from each intervention from the program as a whole,
- Specify the outcomes to be achieved,
- Are comprised of interventions known to effect a particular outcome,
- Coordinate and organize the community to focus on the issue, and
- Are based on a clear plan with goals, objectives and implementation steps" (DHHS 2001).

Different nations take somewhat different approaches to suicide prevention. The WHO guidelines encourage a collaborative approach, with no single agency creating or implementing the strategy. Finland is an example of a country with strong community involvement in development and implementation. The development of the national strategy in the United States, although led by the federal government, has relied on collaboration from numerous organizations and individuals.

"National suicide prevention strategies vary in terms of their target audiences. The *National Strategy* is aimed at the entire population of the U.S. and in this respect is similar to the strategies of Norway, Sweden, and Finland. In contrast, New Zealand and Australia focus exclusively on youth suicide. Finland has also targeted young men for special attention, given their increasing rate of suicide in that country" (SAMHSA 2001).

Rosenhan Study

Rosenhan, David L. (1975). On being sane in insane places. *Science* 179: 250–258.

In 1975 *Science* published a report on an experiment by David L. Rosenhan, a professor at Stanford University. He had eight normal people apply for admission to psychiatric hospitals in five states on the east and west coasts. Each person complained of hearing voices and was admitted to the hospitals. Once admitted, the people were never found out. Even the pseudopatients' constant note taking did not make the staff wary but was interpreted as an aspect of their pathology. Only actual patients detected the deception. Eventually all the pseudopatients were discharged as in remission. The antipsychiatry movement used this study.

World Federation for Mental Health

Available at http://www.wfmh.org/aboutus/mission.html.

"The World Federation for Mental Health (WFMH) was founded in 1948 to advance, among all peoples and nations, the prevention of mental and emotional disorders, the proper treatment and care of those with such disorders, and the promotion of mental health."

The World Federation of Mental Health has members in 112 countries. The organization provides worldwide advocacy and public organization. The federation's membership includes consumers, family members, mental health professionals, and concerned citizens. The organization collaborates with governmental and nongovernmental organizations and is a consulting organization to the United Nations.

Following are the WFMH's mission, vision, and goals:

"MISSION: The mission of the World Federation for Mental Health is to promote, among all people and nations, the highest possible level of mental health in its broadest biological, medical, educational, and social aspects." "VISION: The World Federation envisions a world in which mental health is a priority for all people. Public policies and programs reflect the crucial importance of mental health in the lives of individuals, families and communities, and in the political and economic stability of the world. The interdependence of mental and physical health within the social environment is fully recognized. Serious and effective programs are focused on research, training and services for promotion of mental health and optimal functioning, prevention of disorders, and care and treatment of those with mental health problems throughout the life cycle. Those who experience mental, neurological and psychological disorders are understood and accepted, and treated equitably in all aspects of community life.

Goals

- "To heighten public awareness about the importance of mental health, and to gain understanding and improve attitudes about mental disorders.
- To promote mental health and optimal functioning.
- To prevent mental, neurological and psychosocial disorders.
- To improve the care and treatment of those with mental, neurological and psychosocial disorders."

Information on mental health programs and activities around the world can be found on the World Federation for Mental Health's Web site. It is a good site for the most current information.

World Health Report 2001—*Mental Health: New Understanding, New Hope*

Author: World Health Organization, Geneva, 2001.

Available at http://www.who.int/whr/2001/overview/wn/index.html.

The World Health Organization (WHO) publishes special reports yearly. The 2001 special report was the first time WHO published a special report on mental health. The report was published to raise awareness by professionals and the public about "the real burden of mental disorders and their costs in human, social and economic terms." The report also sought to reduce the barriers of stigma, inadequate services, and discrimination. The report was a "comprehensive review of what is known about the current and future burden of disorders, and the principal contributing factors. It examines the scope of prevention and the availability of, and obstacles to, treatment. It deals in detail with service provision and planning; and it concludes with a set of far-reaching recommendations that can be adapted by every country according to its needs and its resources."

The report makes ten recommendations for action:

1. "Provide treatment in primary care

The management and treatment of mental disorders in primary care is a fundamental step which enables the largest number of people to get easier and faster access to services. It needs to be recognized that many are already seeking help at this level. This not only gives better care; it cuts wastage resulting from unnecessary investigations and inappropriate and non-specific treatments. For this to happen, however, general health personnel need to be trained in the essential skills of mental health care. Such training ensures the best use of available knowledge for the largest number of people and makes possible the immediate application of interventions. Mental health should therefore be included in training curricula, with refresher courses to improve the effectiveness of the management of mental disorders in general health services.

2. Make psychotropic drugs available

Essential psychotropic drugs should be provided and made constantly available at all levels of health care. These medicines should be included in every country's essential drugs list, and the best drugs to treat conditions should be made available whenever possible. In some countries, this may require enabling legislation changes. These drugs can ameliorate symptoms, reduce disability, shorten the course of many disorders, and prevent relapse. They often provide the first-line treatment, especially in situations where psychosocial interventions and highly skilled professionals are unavailable.

3. Give care in the community

Community care has a better effect than institutional treatment on the outcome and quality of life of individuals with chronic mental disorders. Shifting patients from mental hospitals to care in the community is also cost-effective and respects human rights. Mental health services should therefore be provided in the community, with the use of all available resources. Community-based services can lead to early intervention and limit the stigma of taking treatment. Large custodial mental hospitals should be replaced by community care facilities, backed by general hospital psychiatric beds and home care support, which meet all the needs of the ill that were the responsibility of those hospitals. This shift towards community care requires health workers and rehabilitation services to be available at community level, along with the provision of crisis support, protected housing, and sheltered employment.

4. Educate the public

Public education and awareness campaigns on mental health should be launched in all countries. The main goal is to reduce barriers to treatment and care by increasing awareness of the frequency of mental disorders, their treatability, the recovery process and the human rights of people with mental disorders. The care choices available and their benefits should be widely disseminated so that responses from the general population, professionals, media, policy-makers and politicians reflect the best available knowledge. This is already a priority for a number of countries, and national and international organizations. Well-planned public awareness and education campaigns can reduce stigma and discrimination, increase the use of mental health services, and bring mental and physical health care closer to each other.

5. Involve communities, families and consumers

Communities, families and consumers should be included in the development and decision-making of policies, program, and services. This should lead to services being better tailored to people's needs and better used. In addition, interventions should take account of age, sex, culture and social conditions, so as to meet the needs of people with mental disorders and their families.

6. Establish national policies, programs and legislation

Mental health policy, programs and legislation are necessary steps for significant and sustained action. These should be based on current knowledge and human rights considerations. Most countries need to increase their budgets for mental health programs from existing low levels. Some countries that have recently developed or revised their policy and legislation have made progress in implementing their mental health care programs. Mental health reforms should be part of the larger health system reforms. Health insurance schemes should not discriminate against persons with mental disorders, in order to give wider access to treatment and to reduce burdens of care.

7. Develop human resources

Most developing countries need to increase and improve training of mental health professionals, who will provide specialized care as well as support the primary health care programs. Most developing countries lack an adequate number of such specialists to staff mental health services. Once trained, these professionals should be encouraged to remain in their country in positions that make the best use of their skills. This human resource development is especially necessary for countries with few resources at present. Though primary care provides the most useful setting for initial care, specialists are needed to provide a wider range of services. Specialist mental health care teams ideally should include medical and non-medical professionals, such as psychiatrists, clinical psychologists, psychiatric nurses, psychiatric social workers and occupational therapists, who can work together towards the total care and integration of patients in the community.

8. Link with other sectors

Sectors other than health, such as education, labor, welfare, and law, and nongovernmental organizations should be involved in improving the mental health of communities. Nongovernmental organizations should be much more proactive, with better-defined roles, and should be encouraged to give greater support to local initiatives.

9. Monitor community mental health

The mental health of communities should be monitored by including mental health indicators in health information and reporting systems. The indices should include both the numbers of individuals with mental disorders and the quality of their care, as well as some more general measures of the mental health of communities. Such monitoring helps to determine trends and to detect mental health changes resulting from external events, such as disasters. Monitoring is necessary to assess the effectiveness of mental health prevention and treatment programs, and it also strengthens arguments for the provision of more resources. New indicators for the mental health of communities are necessary.

10. Support more research

More research into biological and psychosocial aspects of mental health is needed in order to increase the understanding of mental disorders and to develop more effective interventions. Such research should be carried out on a wide international basis to understand variations across communities and to learn more about factors that influence the cause, course and outcome of mental disorders. Building research capacity in developing countries is an urgent need."

The report recognizes that different countries have varying levels of mental health resources, so the report recommends three different scenarios for action based on three different levels of resources. "Scenario A, for example, applies to economically poorer countries where such resources are completely absent or very limited. Even in such cases, specific actions such as training of all personnel, making essential drugs available at all health facilities, and moving the mentally ill out of prisons, can be applied. For countries with modest levels of resources, Scenario B suggests, among other actions, the closure of custodial mental hospitals and steps towards integrating mental health care into general health care. Scenario C, for those countries with most resources, proposes improvements in the management of mental disorders in primary health care, easier access to newer drugs, and community care facilities offering 100% coverage."

These are just some of the reports and their findings that are available to assist research and understanding regarding mental health policy and services. The next chapter will address some of the major health laws and court cases that impact mental health policy and services.

References

Cook, J. A., S. Terrell, and J. A. Jonikas (2004). *Promoting Self-Determination for Individuals with Psychiatric Disabilities through Self-Directed Services: A Look at Federal, State, and Public Systems as Sources of Cash-Outs and other Fiscal Expansion Opportunities.* Paper prepared for the SAMSHA Consumer Direction Summit, March 2004.

Dougherty, R. H. (2003). Consumer Directed Healthcare: The Next Trend? *Behavioral Healthcare Tomorrow* 12: 21.

Murray, J. L. and A. D. Lopez Eds. (1996). *The Global Burden of Disease.* Geneva: World Health Organization, Harvard School of Public Health, and World Bank.

National Council on Disabilities (2000). From Privileges to Rights: People Labeled with Psychiatric Disabilities Speak for Themselves. Washington, DC: National Council on Disabilities.

Peters et al (1998). National Vital Statistics Reports: *Deaths: Final Data for 1996.* Atlanta: Centers for Disease Control.

Surgeon General (2000). *Mental Health: A Report of the Surgeon General.* Washington, DC: Department of Health and Human Services.

U. S. Department of Health and Human Services, Public Health Service (2001). *National Strategy for Suicide Prevention: Goals and Objectives for Action.* Rockville, MD: U. S. Department of Health and Human Services.

8

Legislation and Court Cases

Mental health law has focused on the conflict between individual autonomy and government intervention to provide treatment or protection. Before the 1960s, mental health law consisted of a few cases involving negligence in the provision of care and the releasing of patients prematurely resulting in death, usually by suicide. Most state civil commitment laws allowed commitment for an indefinite period of time based on a certification by a physician that the person was mentally ill. The physician was often a family doctor and not a psychiatrist. Mental health law emerged in the 1960s when people with mental illness were still largely confined in large state institutions. In 1960 Morton Birnbaum, a physician and attorney, advocated that persons confined involuntarily to state psychiatric hospitals had a constitutional right to treatment (Birnbaum 1960). Using tactics developed in the civil rights movement of African Americans, civil rights lawyers argued that mental illness alone should not deprive an individual of his/her civil rights. The federal courts played a dominant role for two decades. Mental health law now plays a significant role in the organization, financing, and delivery of mental health services.

Protection

Government intervention in the lives of the mentally disabled historically has been based on the doctrine of *parens patriae* (Latin: literally, "parent of the country") and the police powers of the state to protect health, safety, welfare, and morals. In England, the monarch as parens patriae was long regarded the general guardian of "infants, idiots, and lunatics" and was responsible for the care and custody of "all persons who had lost their intellects and become . . . incompetent to take care of themselves" (Merrick 1975, 255). The state now assumes that sovereign responsibility to protect and care for the mentally disabled. So the state may within constitutional limits enact provisions for the protection of the mentally disabled. Such provisions usually are procedures for involuntary commitment to an institution for treatment, care, or custody. Generally involuntary commitment is founded on dangerousness to self or others, including inability to care for oneself. Persons facing loss of liberty by civil commitment have not always had due process safeguards, but these safeguards have developed through legislation and court cases.

Common law extension of procedural safeguards was slow, as the traditional approach was that of the "benevolent benefactor," the view of the state as not punishing but protecting and caring for the individual (Merrick 1975). A major erosion of that view occurred with the extension of some protections to juveniles in *In re Gault* (387 U.S. 1; 18 L. Ed. 2d 527; 87 S. Ct. 1428 1967). That case was followed by commitment cases involving persons with mental illness and mental retardation.

Commitment

In *Covington v. Harris* (419 F. 2d 616 D.C. 1969), the courts held that the Constitution requires a demonstration that there are no appropriate less restrictive alternatives before a mentally disabled person can be involuntarily hospitalized.

In *Dixon v. Attorney General* (325 F. Supp. 966 M.D. Pa. 1971), a Pennsylvania federal court detailed procedural safeguards affecting involuntary commitment procedures for the mentally retarded. These included notice, a full hearing, counsel, and a mental examination by an independent expert, who might be appointed by the court if the person was indigent. At the hearing, the allegedly mentally retarded person could present evidence, subpoena witnesses and documents, and cross-examine witnesses. The standard for commitment was unequivocal and convincing evidence that the person posed a present threat of serious physical harm to self or others.

The order of the district court in *Wyatt v. Stickney* (344 F. Supp. 387 M.D. Ala. 1972) banned commitment to an institution unless it was first determined that residence in the institution was the least restrictive habilitation setting for that individual.

The conditions of state hospitals played a role in the filing of cases. In *State ex rel. Hawks v. Lazaro* (202 S. E. 2d 109 1974) the West Virginia Supreme Court ruled that its state commitment statute was unconstitutional and found the state psychiatric hospital to be in a condition of "Dickensian squalor." (*State ex rel. Hawks v. Lazaro,* 1974 120). In a 1974 South Dakota case, *Schneider v. Rodeck* (S.D. Cir. Ct. May 9, 1974), the court ruled involuntary commitment statutes must provide for adequate notice, effective counsel, and clear definitions and standards. The court ordered "controlled release" (release at the earliest practical date to the least restrictive treatment) as a solution to remedying the violation and still protecting society. Also in 1974, a consent decree was entered in *Jobes v. Michigan Department of Mental Health* (No. 74-004-130 D.C. Mich. Cir. Ct., Wayne Co. Feb. 26, 1974). That agreement stipulated that no mentally ill, epileptic, or mentally retarded child could be received by the Department of Mental Health in Michigan without a judicial hearing. In addition, the 3,000 children already committed by their parents or guardians were to receive hearings to determine their admission status. In the same year Michigan law (Michigan Compiled Laws Annotated 330.1511) was changed to give minors of the age thirteen or older the right to invoke a court hearing to protest parental placement decisions. Treatment personnel were required to inform all recipients of services of their rights (Michigan Compiled Laws Annotated 330.1706).

One of the most extensive cases involving involuntary commitment was the Wisconsin case, *Lessard v. Schmidt* (421 U.S. 957 1975). It was the seminal case in drastically limiting the state's power to commit patients on parens patriae grounds.

In 1971, after a reported suicide attempt, the police picked up Alberta Lessard, and a judge committed her for treatment. She was diagnosed with paranoid schizophrenia. Lessard contacted the federally funded Milwaukee Legal Services, which brought a class action suit on behalf of all adults held on the basis of Wisconsin's involuntary civil commitment statutes. The suit challenged both the grounds on which the state committed mentally ill persons (as being overly vague) and the procedures that were used. Wisconsin, like most other states at that time, had a loose commitment statute, providing that a mentally ill individual could be committed if "a proper subject for custody and treatment." A mentally ill person was defined as someone who "requires care and treatment for his own welfare, or the welfare of others in the community." Procedures were also relaxed, with judges relying heavily on psychiatric opinion and the patient having little if any input and no right to counsel. In the case of Alberta Lessard, the judge had appointed a guardian to act in her best interests, but in the opinion they wrote on the case, the Lessard judges complained that he did not serve an adversary role. After less than a month in the hospital, Lessard was sent to an outpatient facility.

The principles decided in that case were that no significant deprivation of liberty can be justified without a hearing. Procedural safeguards required are adequate notice, a full hearing, right to a jury trial, notice of the grounds for detention, standard on which one may be detained, and the names of examining physicians. The standard for commitment must be an overt action dangerous to others, as proven beyond a reasonable doubt.

Lessard v. Schmidt revolutionized mental health law. It was the beginning of the end for the broad commitment statutes that were then the general rule, as state after state followed Wisconsin in sharply constricting or all but abandoning the traditional parens patriae grounds for commitment in favor of strictly focusing on the police power. Involuntary civil commitment ceased to be viewed as a primarily medical decision (albeit one authorized by a court). It began to be viewed as a quasi-criminal proceeding, with the individual to be accorded all the procedural protections of the criminal law. From that point on, primarily lawyers, not doctors, have made commitment decisions.

Bartley v. Kremens (402 F. Supp. 1039 E.D. Pa. 1975) was the lead case in "voluntary" commitment of minors. That class action suit was filed on behalf of all citizens eighteen years old or younger who may be committed to facilities for the mentally ill or mentally retarded under Pennsylvania statutory provisions. The Pennsylvania law permitted commitment of juveniles simply upon application of parents or guardians, regardless of the wishes of the juvenile and without a hearing or counsel for the juvenile. The federal court declared the challenged commitment statutes unconstitutional. The court held that due process requirements apply to civil commitment proceedings involving minors. The required safeguards included a probable cause hearing within seventy-two hours from the date of initial detention; a postcommitment hearing within two weeks from initial detention; written notice, including date, time, place, and grounds for the proposed commitment; right to counsel; right to be present at all hearings; a finding of clear and convincing proof of the need for institutionalization; the rights to confront and to cross-examine witnesses, to offer evidence, and to testify. The court said that parents may not waive the constitutional rights of their children.

The Georgia case *J. L. and J. R. v. Parham* (No. 75-163-Mac. M. D. Ga. February 16, 1976) challenged a commitment law similar to the one invalidated in *Bartley*. The court struck down the challenged Georgia "voluntary" admission statute as unconstitutional. The court then ordered the defendants to release the approximately two hundred children involved or initiate recommitment proceedings within sixty days under Georgia laws that had not been found unconstitutional.

In 1979 in *Addington v. Texas* (441 U.S. 418, 1979), the U.S. Supreme Court ruled that civil commitment only requires proof by "clear and convincing evidence," while a criminal finding requires proof "beyond a reasonable doubt." In the same year, children were exempted from the judicial hearing requirements for an adult in *Parham v. J.R.* (442. U.S. 584 1979). Children may be admitted to a state hospital on the application of their parents or guardian or upon application of the state as their acting guardian. These loose commitment standards allowed the widespread commitment of children and adolescents in the 1980s.

Commitment and Criminal Justice

Persons with mental disabilities may be found incompetent to stand trial. In 1968 the Supreme Court of Massachusetts in *Nelson v. Superintendent of Bridgewater State Hospital* (233 N.E. 2d 908 Mass. 1968) enunciated a constitutional right to treatment for persons found incompetent to stand trial. Then in 1972 the U.S. Supreme Court handed down the historic decision, *Jackson v. Indiana* (406 U.S. 715 1972). That case involved a twenty-seven-year-old deaf-mute with a mental age of three to four years. He was accused of taking nine dollars in two larcenies. Although he denied the charges, he was declared incompetent and not allowed to stand trial. He was confined to an Indiana state mental hospital for almost three years. If he had been convicted of the larcenies, the maximum sentence he could have received was six months (*Mental Retardation and the Law* 1975). If his case had not been taken to court, he probably would have remained institutionalized for life.

The case raised constitutional questions of equal protection, due process, right to bail, and right to a speedy trial. The Supreme Court held that requiring Jackson to face a more lenient commitment standard and a more stringent release standard than those not charged with offenses was a deprivation of equal protection and that indefinite commitment of a criminal defendant because of incompetence to stand trial violated due process under the Fourteenth Amendment. The court held that a person who was committed solely because of his incapacity to go to trial cannot be held more than the reasonable period of time necessary to determine whether there is a substantial probability that he will attain the capacity to go to trial in the foreseeable future. If that cannot be shown, he must either be released or the state must move against him under the regular civil commitment procedures. If it is shown he may soon be able to stand trial, his continuing commitment must be justified by progress toward gaining competency to stand trial.

A number of cases patterned after *Jackson* followed. In *State ex rel. Matalik v. Schubert* (57 Wis. 2d 315, 204 N.W. 2d 23 1972) the Wisconsin Supreme Court ruled that an involuntarily committed person cannot be held more than six months. At the end of that time, he must either be released or the state must seek a civil commitment. The following year, in *State ex rel. Haskins v. County Court of Dodge County* (62 Wis. 2d 250, 214 N.W. 2d 575 Supreme Ct. Wis. 1974), the Wisconsin Supreme Court ruled that the state could confine a person found incompetent to stand trial for a total of eighteen months without violating *Jackson*. However, a hearing must be held every six months on the question of competency. In a West Virginia case, *State ex rel. Miller v. Jenkins* (13350 Supreme Ct. of Appeals, W. Va. at Charleston 1974), the Supreme Court of Appeals determined that a person's competency to stand trial must be decided within sixty days of commitment. Someone found to be incompetent could not be held more than six months. The court also ruled a person could not be committed to a mental institution in connection with a criminal prosecution unless there was evidence that the person was dangerous to self or others, and that a severely mentally retarded person must be either civilly committed or released. These cases greatly reduced the power of judges to commit people to institutions for indefinite periods at their own discretion and reduced the power of the administrators of an institution to keep such people in the institution at their own discretion.

Right to Treatment

As mentioned before, in 1960 Dr. Morton Birnbaum, attorney and physician, wrote in the *American Bar Association Journal* that persons involuntarily committed to mental institutions have a right to treatment. From this article grew a movement for requiring treatment for the mentally ill and habilitation for the mentally retarded. This movement was to grow to include the voluntarily committed as well as the involuntarily committed.

The first legal case expressing a right to treatment was *Rouse v. Cameron* (373 F. 2d 451 D.C. Cir 1966). Seventeen-year-old Charles Rouse was arrested in Washington, D.C., for carrying a gun and ammunition. The typical criminal sentence was one year, but Rouse was diagnosed as a sociopath and given an indeterminate sentence at St. Elizabeth's, a mental hospital. After four and a half years, a legal aid lawyer brought the case to the federal district court, where he lost. The appeal was heard before the Washington, D.C., Circuit Court. Judge David Bazelon ordered the case returned to the lower court for examination of the question of whether Rouse had been deprived of his right to treatment. The issue of right to treatment was grounded in Washington, D.C., statutes rather than the U.S. Constitution. *Rouse* was significant for establishing the right of the court to intervene in the internal affairs of institutions. It also brought the issue of a right to treatment into court and triggered a debate among lawyers and psychiatrists about judicial review of adequacy of treatment. As for Charles Rouse, he was released on the grounds that his original insanity pleas had been interposed over his objection.

In 1972 the case *Wyatt v. Stickney* (344 F. Supp. 387 M. D. Ala. 1972) sought in federal court to establish a constitutional right to treatment. The case developed in Alabama in 1970 when some workers at Bryce Hospital for the mentally ill were terminated. The employees hired a lawyer to get their jobs back. He had not heard of *Rouse*, but from his readings on rights of prisoners and wards of the state, he developed a right-to-treatment argument and brought a class action suit. He hoped that the establishment of a right to treatment would force the institution to upgrade care and that that would result in the employees being rehired. The case was brought in the name of Ricky Wyatt, a patient at the hospital, whom the lawyer had never met, and all other persons similarly situated (Goodman 1974).

Bryce had a population of 5,000 persons, including 1,000 people with mental retardation. Most of the population received only custodial care. The issues summarized in the case were as follows: "Is the basis for civil commitment such that the state is required to provide adequate treatment procedures so as not to abridge the constitutional rights of the patients" and "does the Alabama Department of Mental Health provide treatment facilities that conform to minimum acceptable standards?" (*Wyatt v. Stickney*).

Judge Frank Johnson of the federal district court determined that the treatment programs were scientifically and medically inadequate and deprived the patients of their constitutional rights. The court said: "To deprive any citizen of his or her liberty upon the altruistic theory that the confinement is for humane therapeutic reasons and then fail to provide adequate treatment violates the very fundamentals of due process" (*Wyatt v. Stickney,* 388). The court said that, from a constitutional viewpoint, the only justification for civil commitments was treatment. The court also said that failure to provide adequate treatment because of lack of funds for staff or facilities was no justification.

The court gave the state of Alabama six months to improve care. When, at the end of that time, the court was not satisfied with the state's action, the judge invited the U.S. Justice Department to submit an amicus curiae brief. Amicus curiae briefs were also submitted by several organizations, including the American Civil Liberties Union, the American Psychological Association, the American Association on Mental Deficiency, the American Psychiatric Association, the National Association of Retarded Citizens, and the National Association for Mental Health. At this point patients from Searcy Hospital for the mentally ill and residents from Partlow State School and Hospital for the Mentally Retarded had joined the case.

The basic principles established by *Wyatt* were that there is a constitutional right to treatment and that that right requires the following:

1. "An individual treatment program;
2. A humane physical and psychological environment;
3. An adequate and qualified staff;
4. Programs provided in the least restrictive manner possible" (President's Committee on Mental Retardation 1973).

Separate standards were issued by the court for the mentally ill and for the mentally retarded.

The *Wyatt* case was appealed to the Fifth Circuit Court of Appeals. Joined with it on appeal was a 1972 Georgia case, *Burnham v. Department of Public Health of the State of Georgia* (349 F. Supp. 1355 N. D. Ga. 1972) which had drawn conclusions quite opposite to *Wyatt*. The *Burnham* case was filed on behalf of patients and former patients of six Georgia institutions. Violations of the Fifth, Eighth, and Fourteenth Amendments to the U.S. Constitution were charged, and a permanent injunction and declaratory judgment similar to those in *Wyatt* were sought. Judge Sydney Smith of the U.S. District Court for the Northern District of Georgia granted the defendant's motion to dismiss.

"Although Judge Smith recognized that persons committed to Georgia's mental institutions might have a *moral* right to effective treatment, he disagreed with plaintiffs that Georgia was under a *legal* obligation to provide such treatment . . . Primarily he found no legal precedent for a ruling that there is a federal constitutional right to treatment. While Judge Smith was aware of the *Wyatt* decision, he stated, 'This Court respectfully disagrees with the conclusion reached by that court in finding an affirmative federal right to treatment absent a statute so requiring.' Moreover the court interpreted the Eleventh Amendment to prohibit a federal court from requiring state expenditures in an area controlled by state law. Judge Smith further suggested that treatment of involuntary patients in mental institutions is not a 'justiciable issue;'—i.e., not an issue capable of definition and resolution by a court. Finally he indicated that the establishment and policing of *individualized* treatment cannot be undertaken by a court, and should be left to the discretion of the professionals rendering services" (*Burnham v. Department of Public Health of the State of Georgia*, 349 F. Supp. 1355 N. D. Ga. 1972).

For the appeal of the *Wyatt* case, lawyers for Governor George Wallace and the State of Alabama used the identical arguments that had persuaded the Georgia court to dismiss *Burnham*. They argued that a federal court could not tell a state how to allocate its resources, and that only a legislature can decide whether it is more important to provide a pension for elderly having no other income than to provide expensive psychiatric treatment and other services to patients at mental institutions.

While on appeal, the *Wyatt* case received two name changes as changes were made in Alabama Commissioners of Mental Health. The case became *Wyatt v. Aderholt* (503 F. 2d 1305 5th Cir. 1974) and then *Wyatt v. Hardin* (503 F. 1d 1305 5th Cir. 1974). Then on November 8, 1974, the United States Court of Appeals for the Fifth Circuit in part affirmed and in part modified *Wyatt*. Of greatest significance, the court affirmed that the Constitution guarantees civilly committed persons in state institutions for the mentally ill and mentally retarded a right to treatment and habilitation. The court, however, did not specifically uphold the standards written for the Alabama institutions because Alabama did not appeal the specific conditions and standards set by the district court. There was no decision as to whether those standards were the constitutional minimum required.

The court of appeals based its decision on its earlier decision in the case *Donaldson v. O'Connor* (49 3F. 2d 507 5th Cir. 1974), in which a former mental patient sought damages for a fifteen-year confinement without treatment. However, the court of appeals replied to one argument not answered in *Donaldson*. Governor Wallace had challenged the assumption that danger to self or others or inability to care for one's needs were the only allowable justifications for treatment. He suggested that the principal justification lay in the inability of the mentally ill and the mentally retarded to care for themselves, and that the real clients of the institutional system were the families and friends of the mentally disabled. "Obviously, if 'need for care' is a justification for commitment, then it follows that mere provision of custodial care is constitutionally adequate to justify continued confinement" (*Wyatt v. Hardin*). The court said that the care at Alabama hospitals was not even adequate enough to meet the more limited standards of "adequate care." The court went on to reject the need for care and the relief of the families as constitutional justification for civil commitment.

"At stake in the civil commitment context, as we emphasized in *Donaldson* are 'massive curtailments' of individual liberty. Against the sweeping personal interests involved, Governor

Wallace would have us weigh the state's interest, and the interests of the friends and families of the mentally handicapped in having private parties relieved of the 'burden' of caring for the mentally ill. The state interest thus asserted may be, strictly speaking, a 'rational' state interest. But we find it so trivial beside the major personal interests against which it is to be weighted that we cannot possibly accept it as justification for the deprivations of liberty involved. . . . Our express holding in *Donaldson* and here rests on the *quid pro quo* concept of 'rehabilitative treatment,' or, where rehabilitation is impossible, minimally adequate habilitation and care, beyond the subsistence level custodial care that would be provided in a penitentiary" (*Wyatt v. Aderholt*, 1313, 1314).

The court also held that the right to treatment could be implemented through judicially manageable standards. The defendants in *Wyatt* did not appeal the Fifth Circuit's decision to the United States Supreme Court. The *Wyatt* decision stood as a final decree. The *Wyatt* decision established a constitutional right to treatment and habilitation up to the circuit court level, but it did not establish the right at the Supreme Court level.

Three constitutional provisions were used to establish the right to treatment and habilitation. Under the due process clause of the Fourteenth Amendment, fundamental fairness requires that more than mere custodial care is needed to justify loss of liberty. The Eighth Amendment prohibits cruel and unusual punishment, and since civil commitment without treatment and habilitation would amount to punishment for being mentally disabled, the commitment would violate the Eighth Amendment. Finally, many state institutions did not meet the standards of other institutions for fire safety, food sanitation, education, and so on, thus violating the equal protection clause of the Fourteenth Amendment (Ennis and Friedman 1973).

Donaldson v. O'Connor (422 U.S. 563 1975) brought the question of a right to treatment before the Supreme Court. Donaldson was civilly committed in 1955 to a Florida state hospital for the mentally ill. For nearly fifteen years he received little or no psychiatric treatment. A suit brought on his behalf contended that he had a constitutional right to treatment or release and included a damage suit against the hospital and mental health officials for depriving him of his constitutional rights. The case was upheld in the lower courts on the basis that an involuntarily confined mental patient has the right to treatment or release. The Supreme Court, however, focused its decision on the constitutional right to liberty and issued a narrow opinion. The decision in *Donaldson* did not decide whether the state may confine a non-dangerous mentally ill person if it provides treatment or whether a civilly committed mentally ill person who is dangerous to self or others has a constitutional right to treatment. The court noted that adequacy of treatment is a justiciable question, but there was no specific endorsement of the right to treatment. Also there was no disapproval of that right. In a separate concurrence, Chief Justice Burger indicated an unwillingness to recognize a right to treatment, but no other justice joined him. Four days after the *Donaldson* decision, the court refused certiorari which would have allowed it to hear *Burnham,* the Georgia right-to-treatment case. By refusing to decide the right-to-treatment issue directly, the Court left in effect a number of lower court decisions recognizing a constitutional right to treatment, including *Wyatt.*

In the 1980s court decisions became more conservative. In 1982, in *Youngberg v. Romeo* (457 U.S. 307 1982), the Supreme Court ruled that people in institutions had a right to safety and to freedom from unreasonable restraint. But the Court said that rights "would be reviewed by the courts against a standard that presumed the correctness of the judgment of treatment and administrative officials employed by the state: 'in determining whether the State has met its obligations in these respects, decisions made by the appropriate professional are entitled to a presumption of correctness'" (324; quoted in Petrila and Levin 2004, 48).

Right to Refuse Treatment and Informed Consent

The New York Court of Appeals found in 1914 that, absent an emergency, an adult must give informed consent before being treated (*Schloendorff v. Society of New York Hospital* (211 N.Y. 2d 125; 105 N.E 2d 92 1914). A person is competent who can understand the treatment that is proposed, having cognitive functioning and intellectual maturity to understand the risks and benefits. A person lacking competency must have decisions made by a surrogate.

Early on, individuals committed for psychiatric treatment were assumed to be incompetent. Only more recently has the issue of competency been separated from commitment. Now it is understood that a person's competency is separate from the fact of having a mental illness or having been committed to a psychiatric facility. In *Rivers v. Katz* (67 New York 2d 485, 495 N. E. 2d 337 1986), the New York Court of Appeals said that, though medication has positive effects, it may also have undesirable side effects associated with its use. A person can only be forced to take medication in limited circumstances and only after a judicial review. States now recognize that psychiatric patients who are competent can refuse medication and that that decision may be overridden only after a judicial administrative review.

Advocacy

Individuals with mental illnesses and serious emotional disturbances who reside in treatment facilities are among the groups most vulnerable to potential neglect and abuse. To help ensure that individuals receive appropriate care and treatment, each state has a system, designated by the governor, to protect and advocate for the rights of people with mental illnesses.

The Center for Mental Health Systems (CMHS) funds and oversees the Protection and Advocacy for Individuals with Mental Illness (PAIMI) program, which advocates for individuals with mental illnesses. Protection and advocacy services include general information and referrals; investigation of alleged abuse, neglect, and rights violations in facilities; and use of legal, legislative, systemic, and other remedies to correct verified incidents. Anyone with a mental illness who resides in or recently has been discharged from a facility, such as a hospital, group home, homeless shelter, residential treatment center, jail, or prison, may be eligible to receive these services through the PAIMI program.

The Civil Rights Division Special Litigation Section has ongoing work investigating allegations of inadequate care and treatment in public residential facilities (including mental retardation facilities and adult and juvenile correction facilities) under the Civil Rights of Institutionalized Persons Act. Since 1993, the division has investigated mental health services and monitored remedial settlements to improve the mental health services in more than 300 facilities in forty-two states. The efforts of the Department of Justice (DOJ) also include an ongoing Working Group on Mental Health and Crime and a Suicide Prevention Program.

Legislation

The Social Security Act of 1935 and its Amendments

Social Security provides funding for retirement, disability, and income for survivors of an eligible worker. Several programs under the Social Security Act and its amendments affect people with mental disabilities.

Supplemental Security Income (SSI)

Title II of the Social Security Act, Supplemental Security Income (SSI) Disability Benefits, includes benefits for adults and children. SSI provides income support for children and adults with a physical or mental condition that results in marked or severe functional limitations. The physical or mental condition must last or be expected to last for twelve months.

Medicaid

Congress passed Title XIX of the Social Security Act, establishing Medicaid, in 1965. It is a program jointly funded by the federal government and the state governments that provides health care coverage to low-income individuals and families. Many low-income people with mental disabilities are eligible for Medicaid. Eligibility is based on family size and income. Medicaid is the largest program providing medical and health-related services to low-income people. Within broad national guidelines provided by the federal government, each of the states establishes its own eligibility standards, determines the type and duration of services, sets the rate of payment, and administers its program. Under Medicaid, a range of services can be provided, including inpatient hospital care, residential treatment centers, and group homes. Also covered are physician services, prescription drugs, rehabilitation services, case management, and some community-based services. As the states were historically the financers of the state psychiatric hospitals, Congress did not allow federal financing of services in free-standing psychiatric hospitals. Individuals sixty-five and over were allowed coverage regardless of their location of treatment. In 1972 the law was amended to permit federal reimbursement for persons under the age of twenty-two regardless of the location of treatment.

Early and Periodic Screening, Diagnosis and Treatment (EPSDT)

EPSDT is the child health component of the Medicaid program. Under EPSDT, eligible children receive periodic screening services, including physical exams and vision, dental, and hearing screening. They are entitled to medically necessary services for physical and mental illnesses and conditions. They are entitled to those services even if the state does not otherwise cover that service in its Medicaid plan.

State Children's Health Insurance Program (SCHIP)

Title XXI of the Social Security Act, SCHIP provides health care for children who come from working families with incomes too high to qualify for Medicaid, but too low to afford private health insurance. Under SCHIP, the state can choose to provide child health care assistance to low-income, uninsured children through Medicaid expansion, a separate program, or a combination of the two. SCHIP targets low-income children and in most states defines them as under nineteen and living in families with incomes at or below the poverty line. Children eligible for Medicaid must be enrolled in Medicaid and are not eligible for SCHIP. Also, to be eligible for SCHIP, children cannot be covered by other group health insurance. If a state chooses to expand Medicaid eligibility for its SCHIP program, the children who qualify under SCHIP are entitled to EPSDT. If a state chooses to develop a separate state program to cover children, it must include the same benefits as one of several benchmark plans, or have an actuarial value equivalent to any one of those benchmark plans. Plans based on the equivalent actuarial value must include at least 75 percent of the actuarial value in the benchmark plan for mental health and substance abuse.

The Community Mental Health Centers (CMHC) Act of 1963

This act was the most ambitious mental health legislation in the area of service development. It has been amended numerous times in the 1960s and 1970s. The act sought through federal funding to develop a network of community mental health centers. These centers would provide a set of services that would create a network of centers around the country to provide mental health services in the community instead of in the large psychiatric hospitals. They were to provide services to the seriously mentally ill. The legislation aimed to reduce the psychiatric hospital population by 50 percent over twenty years. Unfortunately, there was little integration with other mental health services and the centers were criticized for serving people with less serious mental disorders and poorly serving the seriously mentally ill.

Head Start 1964

Head Start is a federal preschool program designed to provide educational, health, nutritional, and social services, primarily in a classroom setting, to help low-income children begin school ready to learn. Head Start legislation requires that at least 90 percent of these children come from families with incomes at or below the poverty line; at least 10 percent of the enrollment slots in each local program must be available to children with disabilities.

The Rehabilitation Act of 1973

The Rehabilitation Act forbids discrimination against persons with disabilities, including people with mental disabilities. The act covers access to buildings and transportation and nondiscrimination in employment. The act covers people with physical or mental disabilities.

Individuals with Disabilities Education Act 1975

In administering Part B of the Individuals with Disabilities Education Act (IDEA), the Office of Special Education Programs, U.S. Department of Education, helps states carry out their responsibility to provide all children between the ages of three and twenty-one with a free appropriate public education. In the case of children with disabilities, the emphasis is on special education and related services designed to meet their needs and prepare them for employment and independent living. Children with emotional disturbance may be eligible for special education and related services under IDEA. Additionally, some children with attention deficit hyperactivity disorder may receive services, if identified as eligible under one of the thirteen specific IDEA categories of disability. Eligibility is determined by a multidisciplinary team of qualified school professionals and parents, based on a full and individual evaluation of the child. In addition to special education delivered in the least restrictive environment, eligible children may also receive related services required to assist them to benefit from special education, including psychological services and counseling services.

Each public school child who receives special education and related services under IDEA must have an individualized education program (IEP) that details the child's goals, needed special education and services, and where they will be provided, along with other information. For a child whose behavior impedes his/her learning or that of others, the IEP team can consider positive behavioral interventions, strategies, and supports to address that behavior. The IDEA also provides for functional behavior assessments and development of behavioral intervention plans for students who present challenging and disruptive behaviors.

The Omnibus Reconciliation Act of 1981

This act consolidated fifty-seven federal aid programs into nine block grants. The legislation also repealed the Mental Health System Act, which had been passed the year before at the end of the Carter administration and had not had the opportunity to be implemented.

The McKinney Act of 1987

The McKinney Act makes available funding for states to create housing for people who are homeless and are also mentally ill.

The Fair Housing Amendments of 1988

The Fair Housing Amendments eliminate a number of restrictions in housing for people with mental disabilities. A public hearing can no longer be required before a building permit is granted for housing for persons with mental disabilities.

The Americans with Disabilities Act of 1990

The Americans with Disabilities Act (ADA) provides for access to public transportation and to facilities. Title II of the ADA, which forbids discrimination in the provision of public services, specifies that no qualified individual with a disability shall, "by reason of such disability," be excluded from participation in, or be denied the benefits of, a public entity's services, programs, or activities (§12132). ADA also requires that "reasonable accommodation" be made to employment for a person with disabilities who can perform the "essential job functions" of the job. The act applies to persons with both physical and mental disabilities.

The impact of the ADA on treatment of the mentally ill was addressed in *Olmstead, Commissioner, Georgia Department of Human Resources, et al. v. L. C., by Zimring, Guardian Ad Litem and Next Friend, et al.* (527 U.S. 531 June 22, 1999). The case involved two mentally retarded women, one who was also diagnosed with schizophrenia and the other with a personality disorder. Both women were voluntarily admitted to Georgia Regional Hospital at Atlanta (GRH), where they were confined for treatment in a psychiatric unit. Their treatment professionals eventually concluded that each of the women could be cared for appropriately in a community-based program, but the women remained institutionalized. The women filed suit against state officials under 42 U.S.C. § 1983 and Title II of the ADA to gain access to community care. They alleged that the state had violated Title II in failing to place them in a community-based program once their treating professionals determined that such placement was appropriate. The district court ordered their placement in an appropriate community-based treatment program. The state argued that inadequate funding and not discrimination had kept the women from being placed in community care. The court rejected the state's argument. Under Title II, the court concluded, unnecessary institutional segregation constitutes discrimination per se, which cannot be justified by a lack of funding. The court also rejected the state's defense that requiring immediate transfers in such cases would "fundamentally alter" the state's programs. The case was appealed, and the Eleventh Circuit affirmed the district court's judgment, but remanded for reassessment of the state's cost-based defense. The case was appealed to the U.S. Supreme Court as 138 F. 3d 893.

> "Justice Ginsburg delivered the opinion of the Court concluding that, under Title II of the ADA, States are required to place persons with mental disabilities in community settings rather than in institutions when the State's treatment professionals have determined that

community placement is appropriate, the transfer from institutional care to a less restrictive setting is not opposed by the affected individual, and the placement can be reasonably accommodated, taking into account the resources available to the State and the needs of others with mental disabilities" (138 F. 3d 893). The court ruled that keeping someone institutionalized when they did not require institutionalization was discrimination "by reason of . . . disability." In the ADA, Congress had a "more comprehensive view of the concept of discrimination." The ADA required all public entities to refrain from discrimination under §12132, and identified unjustified "segregation" of persons with disabilities as a "for[m] of discrimination," under §§12101(a)(2) and 12101(a)(5).

Justice Ginsburg was joined by Justice O'Connor, Justice Souter, and Justice Breyer. They determined that once the state provides community-based treatment, it may be limited. "The reasonable-modifications regulation asserts 'reasonable modifications' to avoid discrimination. The ADA does not impel States to phase out institutions, placing patients in need of close care at risk. Nor is it the ADA's mission to drive States to move institutionalized patients into an inappropriate setting, such as a homeless shelter, a placement the State proposed, then retracted. Some individuals might need institutional care from time to time to stabilize acute psychiatric symptoms. For others, no placement outside the institution may ever be appropriate. To maintain a range of facilities and to administer services with an even hand, the State must have leeway. If, for example, the State were to demonstrate that it had a comprehensive, effectively working plan for placing qualified persons with mental disabilities in less restrictive settings, and a waiting list that moved at a reasonable pace not controlled by the State's endeavors to keep its institutions fully populated, the reasonable-modifications standard would be met. In such circumstances, a court would have no reason effectively to order displacement of persons at the top of the community-based treatment waiting list by individuals lower down who commenced civil actions." (138 F. 3d 893)

The Mental Health Parity Act of 1996

The Clinton administration advocated for and signed into law the 1996 Mental Health Parity Act (MHPA). In December 1997, the administration issued regulations to take steps to end discrimination in health insurance on the basis of mental illness under MHPA. As of January 1998, the law began requiring health plans to provide the same annual and lifetime spending caps for mental health benefits as they do for medical and surgical benefits. It does not require full parity, nor does it require a health plan to provide mental health coverage. More than half the states also have some form of parity legislation.

References

Addington v. Texas 441 U.S. 418 (1979).

Bartley, et al. v. Kremens, et al. 402 F. Supp. 1039 E. D. Pa. (1975).

Birnbaum, M. (1960). The Right to Treatment. *American Bar Association Journal* 46:499.

Burnham v. Department of Public Health of the State of Georgia 349 F. Supp. 1355 N. D. Ga. (1972).

Covington v. Harris 419 F. 2d 616 D.C. (1969).

Dixon v. Attorney General 325 F. Supp. 966 M.D. Pa. (1971).

Donaldson v. O'Connor 49 3F. 2d 507 5th Cir. (1974).

Ennis, Bruce, and Paul Friedman, eds. (1973). *Legal Rights of the Mentally Handicapped.* Washington, DC: Practicing Law Institute, The Mental Health Law Project.

Friedman, P., and R. L. Beck, eds. *Mental Retardation and the Law: A Report on the Status of Current Court Cases.* (September 1975): 12.

Goodman, Walter (1974). The Constitution v. the Snakepit. *New York Times Magazine*, March 17, 22.

In re Gault 387 U.S. 1 (1967).

J. L. and J. R. v. Parham No. 75-163-Mac. M. D. Ga. (February 16, 1976).

Jackson v. Indiana 406 U.S. 715 (1972).

Jobes v. Michigan Department of Mental Health No. 74-004-130 D.C. Mich. Cir. Ct., Wayne Co. (Feb. 26, 1974).

Lessard v. Schmidt 421 U.S. 957 (1975).

Merrick, Robert A. (1975). The "Crime" of Mental Illness: Extension of

"Criminal" Procedural Safeguards to Involuntary Civil Commitment. *Journal of Criminal Law and Criminology* 66:255.

Nelson v. Superintendent of Bridgewater State Hospital 233 N.E. 2d 908 Mass. (1968).

Olmstead, Commissioner, Georgia Department of Human Resources, et al. v. L. C., by Zimring, Guardian Ad Litem and Next Friend, et al. 527 U.S. 531 (June 22, 1999).

Parham v. J.R. 442 U.S. 584 (1979).

Petrila, John, and Bruce Lubotsky Levin (2004). Mental Disability, Law, Policy, and Service Delivery. In *Mental Health Services: A Public Health Perspective,* ed. Bruce Lubotsky Levin, John Petrila, and Kevin D. Kennessy. 2nd ed. New York: Oxford University Press.

President's Committee on Mental Retardation (1973). *Compendium of Lawsuits Establishing the Legal Rights of Mentally Retarded Citizens.* Washington, DC: Department of Health, Education, and Welfare.

Rivers v. Katz 67 N. Y. 2d 485, 495 N. E. 2d 337 (1986).

Rouse v. Cameon 373 F. 2d 451 D.C. Cir (1966).

Schloendorff v. Society of New York Hospital 211 N.Y. 2d 125; 105 N.E 2d 92 (1914).

Schneider v. Rodeck S.D. Cir. Ct. (May 9, 1974).

State ex rel. Haskins v. County Court of Dodge County 62 Wis. 2d 250, 214 N.W. 2d 575 S. Ct. Wis. (1974).

State ex rel. Hawks v. Lazaro 202 S. E. 2d 109 (1974).

State ex rel. Miller v. Jenkins 13350 Supreme Ct. of Appeals, W. Va. at Charleston (1974).

State ex rel. Matalik v. Schubert 57 Wis. 2d 315, 204 N.W. 2d 23 (1972).

Wyatt v. Aderholt 503 F. 2d 1305 5th Cir. (1974).

Wyatt v. Hardin 503 F. 1d 1305 5th Cir. (1974).

Wyatt v. Stickney 344 F. Supp. 387 M.D. Ala. (1972).

Youngberg v. Romeo 457 U.S. 307 (1982).

9

Organizations

Professional Societies and Nonprofit Issue and Advocacy Associations

Academy for Health Services Research and Health Policy
1801 K Street NW, Suite 701-L
Washington, DC 20006–1201
(202) 292–6700
http://www.academyhealth.org

The academy includes policy analysts, practitioners, and health services researchers. The academy educates the public on health services research and is a nonpartisan resource for health policy and research. Their research and activities include mental health.

American Association for Marriage and Family Therapy
112 South Alfred Street
Alexandria, VA 22314–3061
(703) 838–9808
http://www.aamft.org/index_nm.asp

The American Association for Marriage and Family Therapy was founded in 1942 as a professional association for marriage and family therapy. It represents marriage and family therapists in the United States, Canada, and abroad. The association engages in promotion, research, and education in the field and establishes standards for the profession.

American Psychiatric Association
1000 Wilson Boulevard, Suite 1825
Arlington, VA 22209–3901
(703) 907–7300
http://www.apa@psych.org

The American Psychiatric Association is a medical specialty society composed primarily of medical specialists who are qualified or are in the process of becoming qualified as psychiatrists. The organization has over 35,000 members both in the United States and internationally. The Web site says that the role of the association is to "work together to ensure humane care and effective treatment for all persons with mental disorder, including mental retardation and substance-related disorders. It is the voice and conscience of modern psychiatry. Its vision is a society that has available, accessible quality psychiatric diagnosis and treatment."

American Psychiatric Publishing
1000 Wilson Boulevard, Suite 1825
Arlington, VA 22209–3901
(703) 907–7322, (800) 368–5777
http://www.psychiatryonline.com

American Psychiatric Publishing is a publishing company under the auspices of the American Psychiatric Association. It makes available textbooks, professional texts, and journals.

American Psychological Association
750 First Street, NE
Washington, DC 20002–4242
(202) 336–5500, (800) 374–2721
http://www.apa.org

The American Psychological Association (APA) is a professional and scientific organization with 150,000 members. It represents psychology in the United States and is the largest association of psychologists in the world. APA's mission statement is on its Web site: "The objectives of the American Psychological Association shall be to advance psychology as a science and profession and as a means of promoting health, education and human welfare by:

- The encouragement of psychology in all its branches in the broadest and most liberal manner

- The promotion of research in psychology and the improvement of research methods and conditions
- The improvement of the qualifications and usefulness of psychologists through high standards of ethics, conduct, education, and achievement
- The establishment and maintenance of the highest standards of professional ethics and conduct of the members of the Association
- The increase and diffusion of psychological knowledge through meetings, professional contacts, reports, papers, discussions, and publications thereby to advance scientific interests and inquiry, and the application of research findings to the promotion of health, education, and the public welfare."

American Public Health Association (APHA)
800 I Street NW
Washington, DC 20001–3710
(202) 777–2742
http://www.apha.org

APHA has a membership of over 30,000 health care professionals in a wide range of disciplines. APHA works to promote public health and equity in health care. APHA has a Mental Health Section, which seeks to identify issues that adversely affect the public's mental health and seeks solutions to improve mental health.

Bazelon Center for Mental Health Law
1101 15th Street, NW, Suite 1212
Washington, DC 20005
(202) 467–5730
http://www.bazelon.org

The Judge David L. Bazelon Center for Mental Health Law is the leading legal advocate for people with mental disabilities in the United States. Cases supported by the center have outlawed institutional abuse and established protections against involuntary commitment. The center advocates in Congress as well as in the courts. Its mission statement is on its Web site: "The mission of the Judge David L. Bazelon Center for Mental Health Law is to protect and advance the rights of adults and children who have mental disabilities. The Center envisions an America where people who have mental illnesses or developmental disabilities exercise their own life choices and have access to the resources that enable them to participate fully in their communities."

The Coalition of Voluntary Mental Health Agencies
90 Board Street, 8th Floor
New York, NY 10004
(212) 742–1600
http://www.cvmha.org

Founded in 1972, the coalition serves as an umbrella advocacy organization for the mental health community in the New York area, representing over 100 nonprofit, community-based mental health organizations that serve over 300,000 clients in the New York City area. It is supported through memberships and foundation and government funding for advocacy and assistance projects. The coalition comprises community mental health centers, private voluntary hospitals, outpatient clinics, day and continuing treatment programs, clubhouses, outreach teams, housing and residential programs, intensive case management programs, settlement houses, and alcoholism and substance abuse services.

Federation of Families for Children's Mental Health
1001 King Street, Suite 420
Alexandria, VA 22314
(703) 684–7710
http://www.ffcmh.org

The federation helps children with mental health needs and their families to achieve a better quality of life, providing leadership for a broad network of family-run organizations. The federation advocates at the national level for the rights of children with mental health needs and their families, and it works to transfer knowledge to statewide family organizations, local chapters, and other family-run organizations to encourage advocacy at the state and local levels, sharing new models, new technologies, and new tools.

Mental Health Matters
Get Mental Help, Inc.
P.O. Box 82149
Kenmore, WA 98028

(425) 402–6934
http://www.mental-health-matters.com

Mental Health Matters was founded to supply information and resources to mental health consumers, professionals, students, and supporters. It offers technical briefs on disorders, symptoms and treatment modes, professional peer discussion boards, and public boards for students, consumers, and professionals.

National Alliance for Hispanic Health
1501 Sixteenth Street, NW
Washington, DC 20036
(202) 387–5000
http://www.hispanichealth.org

The National Alliance for Hispanic Health was established in 1973 by a small coalition of mental health providers. According to its Web site, the mission of the alliance is "to improve the health and well being of Hispanics." The alliance engages in advocacy, information, and research involving both mental and physical health. Some of its specific programs regarding mental health involve consumer outreach in the areas of depression and attention deficit hyperactivity disorder.

National Alliance for the Mentally Ill
Colonial Place Three
2107 Wilson Boulevard, Suite 300
Arlington, VA 22201–3942
(703) 524–7600
http://www.nami.org

Founded in 1979, the National Alliance for the Mentally Ill (NAMI) is a nonprofit, grassroots, self-help, support, and advocacy organization. Its membership includes consumers, as well as families and friends of people with severe mental illnesses, including schizophrenia, bipolar disorder, major depression, obsessive-compulsive disorder, severe anxiety disorders, autism, and attention deficit hyperactivity disorder. NAMI has fifty state organizations and thousands of local affiliates. They provide education and support, support increased funding for research, combat stigma, and advocate for health insurance, housing, employment, and rehabilitation. NAMI's mission statement is given on its Web site: "NAMI is dedicated to the eradication of mental illnesses and to the improvement of the quality of life of all whose lives are affected by these diseases."

National Association of Psychiatric Health Systems
701 13th Street, NW, Suite 950
Washington, DC 20005–3903
http://www.naphs.org

The National Association of Psychiatric Health Systems (NAPHS) was founded in 1933. It represents delivery systems for mental health and substance abuse, including inpatient, residential, partial hospitalization, and outpatient programs, as well as management and prevention services. The NAPHS advocates for high-quality mental health and substance abuse care delivery.

KidsPeace: National Center for Kids Overcoming Crisis
5300 KidsPeace Drive
Orefield, Pennsylvania 18069
(800) 854–3123
http://www.kidspeace.org

KidsPeace is a private nonprofit charity serving the behavioral and mental health needs of children and adolescents. It provides specialized residential treatment services plus a range of treatment programs and education services. Bethlehem Iron Company president William Thurston founded KidsPeace in 1882 for children in crisis. By the 1990s, the organization offered the widest services for child treatment in the United States. The name was changed to KidsPeace Corporation in 1992 to reflect its national identity. The organization services thousands of children referred from national and international sources. Services are provided in several states, including KidsPeace Children's Hospital, residential campuses, primary and secondary schools, diagnostic centers, outpatient clinics, therapeutic foster care, day treatment centers, and therapeutic recreation centers.

In 1991, the organization established a panel of leading national child development experts. The organization has conducted national surveys and public service campaigns, and distributed crisis prevention materials nationally and internationally.

National Center for Policy Analysis
601 Pennsylvania Avenue NW, Suite 900, South Building
Washington, DC 20004

(202) 628-6671
http://www.ncpa.org/

Established in 1983, the National Center for Policy Analysis is a nonprofit, nonpartisan public policy research organization. The center seeks to promote private alternatives to public policy. It conducts some research on mental health policy, including research on the issue of parity between physical and mental health.

National Coalition of Mental Health Professionals and Consumers
P.O. Box 438
Commack, NY 11725
(866) 826-2548
http://www.thenationalcoalition.org

The National Coalition of Mental Health Professionals and Consumers is an advocacy group for people with mental illness and for mental health professionals. It works to create a humane and fair environment for people with mental and emotional disorders. It advocates replacing managed care, which it believes negatively impacts patients and professionals.

National Mental Health Association
2001 N. Beauregard Street, 12th Floor
Alexandria, VA 22311
(703) 684-7722
http://www.nmha.org

The National Mental Health Association (NMHA) was established in 1909 by former psychiatric patient, Clifford Beers. It is the oldest and largest nonprofit organization focused on mental health and mental illness in the United States. NMHA has more than 300 associates. NMHA works to reduce barriers to treatment and services and provides education to millions of people about mental illness. This statement of the mission and vision of NMHA is taken from its Web site: Mission: "The National Mental Health Association is dedicated to promoting mental health, preventing mental disorders and achieving victory over mental illness through advocacy, education, research and service." Vision: "The National Mental Health Association envisions a just, humane and healthy society in which all people are accorded respect, dignity, and the opportunity to achieve their full potential free from stigma and prejudice."

RAND Corporation
1700 Main Street
P.O. Box 2138
Santa Monica, CA 90407–2138
(310) 393–0411
http://www.rand.org/

RAND was established during World War II. It is a major non-profit policy research institute. It conducts research on a wide array of public policy issues, including mental health.

Safe Harbor International Guide to the World of Alternative Mental Health
787 W. Woodbury Road, #2
Altadena, CA 91001
(626) 791–7868
http://www.alternativementalhealth.com

Safe Harbor was founded in 1998. According to its Web site, its mission "is to assist and promote non-harmful, alternative (non-psychiatric) methods and Practitioners for helping the mentally disturbed." Its purpose is to provide education and choice regarding alternative mental health practices.

Government Agencies

Agency for Healthcare Research and Policy
540 Gaither Road
Rockville, MD 20850
(301) 427–1364
http://www.ahcpr.gov/

The Agency for Healthcare Research and Policy works to move research findings into better patient care and provides policy makers and health care leaders with information to make major health care decisions.

Centers for Disease Control and Prevention (CDC)
1600 Clifton Road
Atlanta, GA 30333
(404) 639–3534, (800) 311–3435
http://www.cdc.gov/ or www.cdc.gov/mentalhealth/

The CDC one of the thirteen major organizations of the Department of Health and Human Services and is the lead federal agency for protecting the health of the American people. It responds to public health emergencies at home and aboard. It conducts research on health-related issues and publishes reports. It conducts some research and provides some reports on mental health.

Department of Health and Human Services
200 Independence Avenue, SW
Washington, DC 20201
Telephone: (202) 619–0257
http://www.hhs.gov

The Department of Health and Human Services (DHHS) is the umbrella organization for health of the U.S. government. Agencies of DHHS address both physical and mental health.

National Health Information Center
200 Independence Avenue, SW
Washington, DC 20201
http://www.health.gov/nhic

The National Health Information Center (NHIC) was established in 1979 by the Office of Disease Prevention and Health Promotion, Office of Public Health and Science, Office of the Department of Health and Human Services. The NHIC is a health information and referral service to physical and mental health resources.

National Institute of Mental Health
Public Information and Communications Branch
6001 Executive Boulevard, Room 8184, MSC 9663
Bethesda, MD 20892–9663
(301) 443–4513, (866) 615–6464
http://www.nimh.nih.gov

The National Institute of Mental Health (NIMH) was established in 1946 under the National Mental Health Act, which authorized the Surgeon General to improve the mental health of U.S. citizens through research into the causes, diagnosis, and treatment of psychiatric disorders. The National Institute of Mental Health is one of twenty-seven components of the National Institutes of Health, part of the U.S. Department of Health and Human Services and the federal government's main biomedical and behavioral research agency. The FY 2005 budget of NIMH was $1.4 billion. The NIMH mission is given on its Web site:

"To reduce the burden of mental illness and behavioral disorders through research on mind, brain, and behavior. This public health mandate demands that we harness powerful scientific tools to achieve better understanding, treatment, and eventually, prevention of these disabling conditions that affect millions of Americans.

To fulfill its mission, the Institute:

- Conducts research on mental disorders and the underlying basic science of brain and behavior;
- Supports research on these topics at universities and hospitals around the United States;
- Collects, analyzes, and disseminates information on the causes, occurrence, and treatment of mental illnesses;
- Supports the training of more than 1,000 scientists to carry out basic and clinical research; and
- Communicates information to scientists, the public, the news media, and primary care and mental health professionals about mental illnesses, the brain, behavior, mental health, and opportunities and advances in research in these areas."

National Library of Medicine
8600 Rockville Pike
Bethesda, MD 20894
(301) 594–5983, (888) 346–3656
http://www.nlm.nih.gov

The National Library of Medicine is a part of the National Institutes of Health and is the largest medical library in the world.

Substance Abuse and Mental Health Services Administration (SAMHSA)
Room 12–105, Parklawn Building
5600 Fishers Lane
Rockville, MD 20857
(301) 443–4795
http://www.mentalhealth.samhsa.gov

The Substance Abuse and Mental Health Services Administration works to improve the quality and availability of prevention, treatment, and rehabilitation services in order to reduce illness, disability, death, and costs to society from mental illness and substance abuse.

International Organizations

Center for Addiction and Mental Health
250 College Street
Toronto, Ontario M5T 1R8, Canada
(416) 535–8501
http://www.camh.net

The Center for Addiction and Mental Health (CAMH) is Canada's leading mental health and addiction teaching hospital. CAMH provides health promotion, education, research, and clinical practice. Central facilities are located in Toronto, with twenty-six community locations throughout Ontario. It is affiliated with the University of Toronto and is a Pan American Health Organization and World Health Organization Collaborating Center.

International Society for Mental Health Online
P.O. Box 5464
Bradford, MA 01835
(603) 222–6482
http://www.ismho.org

The International Society for Mental Health Online is a nonprofit organization founded in 1997. Its goal is to promote the understanding, use, and development of online technology, information, and communication for the international mental health community. It provides information about online counseling and psychotherapy, as well as online support groups.

MIND
15–19 Broadway
London E15 4BQ, UK
020 8519 2122
http://www.mind.org.uk

MIND is the leading mental health nonprofit in England and Wales. It advocates for persons with mental health problems. It promotes inclusion, influences policy through lobbying and education, and works for the development of quality services. Its values are autonomy, equality, knowledge, participation, and respect.

Pan American Health Organization
525 23rd Street NW
Washington, DC 20037
(202) 974–3000
http://www.paho.org

The Pan American Health Organization (PAHO) is an international public health agency that is part of the United Nations. It serves as the specialized organization for health of the Inter-American system and as the regional office for the Americas of the World Health Organization. It addresses both physical and mental health.

World Fellowship for Schizophrenia and Allied Disorders
124 Merton Street, Suite 507
Toronto, Ontario, M4S 2Z2, Canada
(416) 961–2855
http://www.world-schizophrenia.org/

The World Fellowship is a global organization for people with schizophrenia and allied disorders and their families. It works to increase knowledge and to reduce stigma and discrimination. It encourages humane treatment of people with mental illnesses. It provides support, training, and education. It provides direct services, self-help groups, and workshops. It produces educational materials and arranges conferences. It advocates for appropriate services and better treatment and works to influence government policy. It also manages research funds. It is made up of twenty-two national family organizations that are voting members and more than fifty smaller groups that are associate members.

World Federation for Mental Health
P.O. Box 16810
Alexandria, VA 22302–0810
Fax (703) 519–7548
http://www.wfmn.com

The World Federation for Mental Health was founded in 1948 to advance the prevention of mental and emotional disorders, promote mental health, and advocate for the proper care and treatment of those with mental disorders. The federation has members and contacts in 112 countries. It is the only worldwide grassroots advocacy and public education organization in the mental health field. Its membership consists of mental health professionals, academics, consumers, family members, and concerned citizens. The federation is accredited as a consultant to the United Nations and its specialized agencies. On the Web site of the federation, the following mission statement is provided: "The mission of the World Federation for Mental Health is to promote, among all people and nations, the highest possible level of mental health in its broadest biological, medical, educational, and social aspects."

World Health Organization
Avenue Appia 10
1211 Geneva 27
Switzerland
41 22 791 2111, FAX 41 22 791 3111
http://www.who.int/en

Established in 1948, the World Health Organization is the United Nations' specialized agency for health. It is governed through the World Health Assembly, which represents 192 member states. As the Web site explains, "WHO's objective, as set out in its Constitution, is the attainment by all peoples of the highest possible level of health. Health is defined in WHO's Constitution as a state of complete physical, mental and social well-being and not merely the absence of disease or infirmity."

10

Selected Print and Nonprint Resources

You can reach a wide array of research resources from your own computer and from school or public libraries. Research from many policy and resource organizations is available directly on the Web. This chapter presents some print and Web resources. When researching, go to the reference desk of your library and ask for assistance in finding these and other resources that fit your research topic.

Books and Reports

Books and government reports are still a primary source of information about mental health policy and services. The following are good sources of information on mental health. These are just a sampling of what is available; when you go to your local library, check to see what other materials are available.

Al-Issa, Ihsan, ed. 1995. *Handbook of Culture and Mental Illness: An International Perspective.* Madison, CT: International Universities Press.

The authors of this edited text explore how psychiatric problems are classified, explained, and treated around the world. Mental disorders are covered in some twenty countries from all major regions of the world. Local experts and practitioners write each chapter, examining the psychopathology and culture of each country.

American Psychiatric Association. 2000. *Diagnostic and Statistical Manual DSM-IV TR.* 4th edition. Arlington, VA: American Psychiatric Association.

The current edition is the fourth, with a text revision in 2000. Written by the American Psychiatric Association, the DSM-IV TR describes over two hundred and fifty types of mental disorders, organized into diagnostic categories, and is widely used in diagnosis.

Beam, Alex. 2003. *Gracefully Insane: Life and Death Inside America's Premier Medical Hospital.* New York: Public Affairs.

This book tells the story of the most elite mental hospital in the United States from its founding in 1817 to the present. Among the many famous people who were committed there are Ray Charles, John Nash, and Sylvia Plath. The author interviewed both patients and staff in order to tell their stories. The author also discusses the hopes and failures of psychology and psychotherapy, the evolution of mental attitudes, and the role of economic pressures.

Desjarlais, Robert, Leon Eisenberg, Byron Good, and Arthur Kleinman. 1995. *World Mental Health: Problems and Priorities in Low-Income Countries.* New York: Oxford University Press.

This book comes from a collaboration of experts from some nineteen countries and researchers in the Department of Social Medicine at Harvard Medical School. They examine mental health problems in Latin America, Africa, the Middle East, and Asia. The authors of this book look at the growing burden of mental, behavioral, and social problems in low-income countries. They examine the mental health consequences of violence, dislocation, poverty, and the disenfranchisement of women. They also examine the impact of mental and behavioral problems on development.

Fellin, Phillip A. 1996. *Mental Illness: Policies, Programs, and Services.* Florence, KY: Wadsworth.

Fellin explores mental health policies, programs, and services; the process and politics of mental health policy making; and historical and current policies and services. He covers special demographic and ethnic groups, including women, children and adolescents, and ethnic minorities. Controversial issues related to mental health policy development are also explored.

Gregory, Richard L., ed. 2004. *The Oxford Companion to the Mind.* 2nd edition. New York: Oxford University Press.

First published in 1987, this new edition has over 1,000 brief articles about the mind by numerous contributors. The book covers descriptions of symptoms and syndromes, sensory and extrasensory perception, theories, consciousness, mental illness and disability, memory, imagination, and language. Contributors include psychiatrists, neuroscientists, physicists, philosophers, psychologists, and linguists. The second edition includes current research on topics such as artificial intelligence, brain imaging, consciousness, and artificial life.

Jacobson, Nora. 2004. *In Recovery: The Making of Mental Health Policy.* Memphis, TN: Vanderbilt University Press.

Since the 1990s, policy makers and providers have been promoting the goal of recovery. Jacobson analyzes what recovery has come to mean as definitions have moved on from the original definition of symptom abatement or a return to a normal state of health. She examines recovery as evidence, recovery as experience, recovery as policy, and recovery as politics, while drawing extensively on her research in Wisconsin, a state with a long history of innovation.

Kemp, Donna R. 1994. *Biomedical Policy and Mental Health.* Westport, CT: Praeger.

This book presents an overview of the relationship between biomedical policy and mental health. It explores the policy issues of a broad array of biomedical research and technology that impacts mental health policy and examines how the very conducting of biomedical research and the use of its technology have implications for the mental health of people. Both negative and positive consequences are possible for mental health policy and for the mental health of human beings as the result of biomedical research and technology. The book does not attempt to provide an in-depth analysis of any particular biomedical research or technology, nor does it explore in detail the public policy or ethics debate of biomedical decision making. Research and technology are used to illustrate their effects on mental health policy and their implications for the mental health of people. The book aims to point out the need to recognize the mental health consequences of the use of biomedical research and technology for all people exposed to those processes.

Kemp, Donna R. 1994. *Mental Health in the Workplace: An Employer's and Manager's Guide.* Westport, CT: Quorum Books.

This book examines how mental health issues impact the workplace and explores ways to create more mentally healthy work environments. The author shows how a mentally healthy workplace can enhance productivity, satisfaction, attendance, and longevity in employment and how companies should comply with federal laws, including the Americans with Disabilities Act. The author points out the importance of mental health in the selection, management, and retention of employees and addresses issues such as violence in the workplace and the effects of corporate culture. She also shows the extent to which mental health plays a role in physical health problems and the cost of inappropriately focusing on physical health care when the underlying issues are mental and emotional health and lifestyle.

Kemp, Donna R., ed. 1993. *International Handbook on Mental Health Policy.* Westport, CT: Greenwood Press.

This book is a comparative look at mental health policy in twenty countries. Each chapter covers the extent of the problems, mental health history, current policy, organization and services, mental health personnel and treatment, public policy process, and special issues, including the mentally ill and the mentally retarded, the mentally ill and substance abuse, the mentally disordered offender, deinstitutionalization, and funding issues.

Levin, Bruce Lubotsky, John Petrila, and Kevin D. Hennessy, eds. 2004. *Mental Health Services: A Public Health Perspective.* 2nd edition. New York: Oxford University Press.

This book integrates information from research in public policy, service systems, treatment methods, and epidemiology to define a public health framework for advancing mental health systems nationally and locally. The book covers delivery and financing of mental health and substance abuse services, state mental health systems, recovery, informatics, psychopharmacology, and the needs of special populations, including those with co-occurring mental and addictive disorders and those in the criminal justice system who have mental disorders.

Lyons, John S. 2004. *Dressing the Emperor: Improving Our Children's Public Mental Health System.* Westport, CT: Greenwood Publishing Group.

Lyons reviews the state of the children's mental health system in the United States and explains why improvements are inconsistent or nonexistent. He includes a brief and inclusive history of the system. He then describes two strategies to change and improve the system. He urges greater investment in employment, housing, child care, and other services to prevent mental illness and then recommends an approach for greater rationality and accountability in service delivery. He recommends using the "Total Clinical Outcomes Management" strategy to use research methodology to guide clinical decision making and program management.

Mechanic, David. 1998. *Mental Health and Social Policy: The Emergence of Managed Care.* Boston: Allyn and Bacon.

David Mechanic examines the role of managed care in mental health and raises questions about the role managed care will play and its accountability. Issues addressed in the book include policy, research, community treatment, case management, epidemiology, and definitions of mental disorders.

Morrall, Peter, ed. 2004. *Mental Health Policy: International Perspectives.* New York: Taylor and Francis.

The authors review mental health policies across the world and their relationship to mentally disordered people. Ten case studies are presented, each written by an expert in mental health policy within that country.

National Institute of Mental Health. *NIMH Study to Guide Treatment Choices for Schizophrenia.* Washington, DC: Government Printing Office. http://www.nimh.nih.gov/press/catie_release.cfm.

The National Institute of Mental Health conducted a study of the effects of five medications for the treatment of schizophrenia to determine whether newer medications were more effective than older medications, and this publication reports on the results of the study. The study included one older medication introduced in the 1950s and four newer medications introduced in the 1990s. Side effects may cause patients to discontinue use of a medication or change treatments. The study found that newer medications did not seem to have any substantial advantage over older medication.

Porter, Roy. 2003. *Madness, a Brief History.* New York: Oxford University Press.

Porter recounts the history of mental illness from antiquity to modern times. He traces changes in attitude over the centuries and discusses various attempts to manage and treat people with mental illness. He includes the development of the patients' rights movement, the development of psychopharmacology, and the rise of psychotherapies.

Poussaint, Alvin F., and Amy Alexander. 2001. *Lay My Burden Down: Suicide and the Mental Health Crisis Among African-Americans.* Boston: Beacon Press.

The authors use interviews and analysis to examine the psychological impact of slavery and persistent racism, including suicide. They examine the social, cultural, and historical factors that make it difficult for blacks to seek health care and discuss ways to bring about change.

President's New Freedom Commission on Mental Health. 2003. *Achieving the Promise: Transforming Mental Health Care in America.* Washington, DC: Government Printing Office. http://www.mentalhealthcommission.gov/reports/Finalreport/FullReport.htm.

The report finds that Americans are living without proper health care and treatment for mental illness. The problem is the manner in which the community-based mental health system has evolved over the past fifty years. The goals of the commission are to help Americans understand the importance of mental health to overall health; to make mental health care family driven; to eliminate disparities in mental health services; to make screening, assessment, and referral expected services; and to make the newest forms of treatment available to those with mental illness.

Reisner, Ralph, Christopher Slobogin, and Arti Rai. 2004. *Law and the Mental Health System, Civil and Criminal Aspects*. Stanford, CT: Foundation.

This book interprets legal doctrine relating to mental health and addresses the relationship between society and the mentally disabled. Relevant empirical and clinical literature is provided.

Rochefort, David, ed. 1989. *Handbook on Mental Health Policy in the United States*. Westport, CT: Greenwood Press.

The authors examine policy issues arising from dwindling budgets and expanding needs. They also explore affected populations and the programs and facilities that serve them. Subject areas include the mental health service system, epidemiological analyses of mental health problems, policy development, the community mental health movement, financial and administrative issues, and future trends.

Smith, Steven R., and Robert G. Meyer. 1987. *Law, Behavior, and Mental Health: Policy and Practice*. New York: New York University Press.

A lawyer and a clinical psychologist address malpractice, confidentiality, the right to treatment and the right to refuse treatment, emotional injuries, competency to stand trial, jury selection, punishment, child custody and child abuse, and violent behavior.

Substance Abuse and Mental Health Services Administration (SAMHSA). 2005. *National Survey of Substance Abuse Treatment Services*. Washington, DC: Government Publications. http://www.oas.samhsa.gov/NSDUH.htm#NSDUHinfo.

This annual report released by SAMHSA shows an increase in drug abuse treatment and a decrease in alcohol abuse treatment. The report also shows that, in general, clients are not being treated for both drug and alcohol abuse, but are being treated for either one or the other. The report contains data on the location, characteristics, and use of drug and alcohol facilities throughout the United States and breaks down statistics on the types of drugs being treated at those facilities. Also included are the proportions of people in treatment for both drug and alcohol abuse.

Torrey, E. Fuller, and Judy Miller. 2002. *The Invisible Plague: The Rise of Mental Illness from 1750 to the Present.* New Brunswick, NJ: Rutgers University Press.

The authors examine the records on mental illness and look at the history of diagnosis and treatment in the United States, England, Ireland, and Canada over a two-hundred-and-fifty-year period. They examine quantitative and qualitative data. They encourage open-minded research into mental illness because so much is still unknown.

Walrond-Skinner, Sue. 1986. *Dictionary of Psychotherapy.* Philadelphia: Routledge and Kegan Paul.

This reference work covers the major theorists, theories, and approaches in the field of psychotherapy. Both classical and modern approaches are covered in the more than 800 entries.

World Federation for Mental Health. *Without Boundaries—Challenges and Hopes for Living with ADHD: An International Survey.* http://www.wfmh.com/publications/without_boundaries.htm.

WFMH spearheaded an international survey of parents with children who suffer from ADHD. Seven countries were included in the survey: Australia, Germany, Italy, Mexico, the Netherlands, Spain, and the United States. The study found that the average time for a child to be diagnosed with ADHD was over two years. This is a long time in the social and academic development of a child. Parents were stressed by their child with ADHD, as well as worried that their child would be left out of social activities and have a hard time at school. Once diagnosed correctly, treatment was found to improve schoolwork, increase social activities, and relieve stress on parents and family members. If parents believe their children are struggling with ADHD, it is important that they be tested and treatment started sooner than later.

World Health Organization (WHO). 2006. *Mental Health Evidence and Research.* Geneva: WHO. http://www.who.int/mental_health/evidence/en/.

At this address, under the general heading "Recent Publications," the reader will find the three useful documents, described below.

"Mental Health Promotion: Case Studies," 2004.

Each case study is focused on a program to promote mental health in countries ranging from Australia to Denmark, China to the United States. Programs were studied for children coping with parents with mental disorders, depression assistance programs, mental health programs in schools, countering stigmas and discrimination associated with mental illness, and public awareness activities.

"Prevention of Mental Disorders," 2004.

This document discusses the need to make the prevention of mental illness a public health priority. Effective preventative tools and programs should be made available to everyone, everywhere. There is a special need to be sensitive to culture and resources across the countries. The protection of human rights is central to the prevention of mental disorders.

"Promoting Mental Health: Summary Report: Concepts, Emerging Evidence, Practice," 2004.

The essential conclusions presented in this summary report are as follows: Mental health hinges on socioeconomic and environmental factors, and is vital for individuals, families, and societies. Effective public health programs that protect basic civil, political, economic, social, and cultural rights can improve maintenance of mental health. Everyone should be involved and interested in the promotion of mental health.

World Health Organization. 2005. *Mental Health Policy, Plans and Programmes.* Geneva: WHO.

The World Health Organization lays out the steps for developing policies, plans, and programs and describes how to implement them. Specific examples from countries around the world are used to illustrate the process of development. WHO believes that an explicit mental health policy is an essential and powerful tool for a mental health agency, and when properly formulated and implemented, policy can have a significant impact on persons with mental illness.

Wright, Rogers H., and Nicholas A. Cummings. 2005. *Destructive Trends in Mental Health: The Well-Intentioned Path to Harm.* Philadelphia: Routledge.

This book makes an argument for a reevaluation of the policies of professional mental health organizations, of delivery of care by practitioners, and of the needs of the public. The essays look at the usurpation of psychology by political correctness, sensitivity and diversity movements, the economic decline in funding for psychology treatment, and the ideological politicization of the profession in regard to scientific research and practice.

Electronic Library-Based Resources

Electronic library-based resources have extensive full-text content that can be accessed by subject. You can access many of these resources at public and school libraries with your library card number or via the Internet from any location.

EBSCO MasterFILE Premier is designed for public libraries and provides full text for over 1,900 general periodicals. Many disciplines are covered, including health, general science, and business. Indexing and abstracts for more than 2,500 publications are provided as well as full text. The database may be searched by keyword or by subject heading.

Health and Wellness Resource Center is a full-text resource with information from journals, medical encyclopedias, and pamphlets.

Medline Plus (http://www.MedlinePlus.gov) is a government-sponsored consumer health site. It can be reached directly from the Web as well as from many library Web sites. The site belongs to the world's largest and most important medical library; the National Library of Medicine. The site contains medical as well psychological and neuropsychological citations.

NewsBank NewsFile provides a full-text database from 1991 to the present of news articles covering health, science, government, social, and other events from over 5,000 U.S. national and regional newspapers, broadcasts, and wire services. The database is searchable by subject headings and keyword.

PsychCrawler is a search engine of the American Psychological Association. It was created to provide quick access to quality content in the field of psychology. The site is currently under rapid development and is adding many new sites.

Psychlit, also referred to as Psycho/Info, is a database and search engine that provides access to publications in psychology, psychiatry, neuropsychology, and neuroscience. It primarily contains journal articles and is fairly easy to use. It should be a primary source for research on mental health and mental illness.

SIRS Publishing Inc. provides electronic and print articles of current issues of interest. Articles are selected for relevance, conciseness, and accuracy. Articles are also selected to provide a balanced view. The audience is people of high-school age.

Health Research and Health Care Policy Organization Web Sites

The following are some of the relevant organizations and issue-oriented Web sites where information on mental health policy, services, and issues may be found.

National Institute of Mental Health (NIMH). http://www.nimh.nih.gov.

The National Institute of Mental Health is the lead government agency for mental health research in the public sector.

Some other government agencies where useful information can be found are the following:

Health Care Financing Administration (HCFA). http://www.hcfa.gov/init/children.htm.

Health Resources and Services Administration (HRSA). http://www.bphc.hrsa.gov and http://www.hrsa.dhhs.gov/childhealth.

Office of Special Education Programs (OSEP). http://www.ed.gov/offices/OSERS/OSEP/index.html.

Some other useful organizations for information on mental health are the following:

OSEP Technical Assistance Center on Positive Behavior Interventions and Supports. http://www.pbis.org.

Center for Effective Collaboration and Practice. http://www.air.org/cecp/.

National Center on Education, Disability, and Juvenile Justice. http://www.edjj.org.

In addition, there is the *Mental Health Directory*, a reference listing of organized mental health services available in the United States, arranged by state, city, and type of service. The directory identifies more than 6,500 mental health organizations and 13,000 service sites. An electronic version can be found on Substance Abuse and Mental Health Services Administration's (SAMHSA's) National Mental Health Information Center Web site at http://www.mentalhealth.samhsa.gov/.

Mental Health Services Locator. http://www.mentalhealth.samhsa.gov/databases/.

The Locator provides comprehensive information about mental health services and resources. You can access this information in several ways by selecting a state or U.S. territory from a map or drop-down menu.

As the home page informs us, "You can go to State Mental Health Resources and find the following:

- Mental Health Facilities Locator. A searchable directory of mental health treatment facilities and support services.
- Mental Health Services Directory. A list of consumer, family, and advocacy organizations that provides comprehensive information about these mental health resources.
- Hispanic Youth Violence Prevention Services. A searchable directory of Hispanic youth violence prevention services.
- State Resource Guide. A summary of national and State organizations that provide professional advocacy protection, family support programs, financing information, and self-help groups.
- State Suicide Prevention Programs. A list of available State-funded suicide prevention plans and programs.
- Substance Abuse Treatment Facility Locator. A searchable directory of nationwide drug and alcohol treatment centers.

- CMHS Grantees. Public and private organizations that receive funding from the Center for Mental Health Services to provide mental health services.
- State Statistics. State-based data on the incidence of mental illness among children and adults, the utilization of inpatient services, and the amount of State spending on mental health."

Glossary

Access Ability of persons needing mental health services to obtain timely and appropriate care.

Acute care Short-term, often intensive, care; can include inpatient care.

Advance directive A legal document signed by a person with mental illness that sets out who they want to be responsible for decision making for them if they are not competent.

Advocacy System to protect the rights of persons with mental illness.

Antipsychiatry movement A movement that declares that mental illness does not exist and believes such labeling is simply a form of social control of those who are different.

APA The American Psychiatric Association; a professional association of psychiatrists.

CDC Centers for Disease Control; a federal agency that conducts research and surveillance and provides grants for research on acute, chronic, and infectious disease.

Chronic disease/disability Medical condition that persists over time. Can be treated outpatient or inpatient.

Committee for Mental Hygiene Founded by Clifford Beers in 1909, it encouraged citizen involvement, prevention of hospitalization, and aftercare.

Community-based services Services provided in the community for people with mental disabilities.

Community Mental Health Centers Outpatient facilities developed with federal funding from the Community Mental Health Centers Act of 1963 that provided diagnosis, treatment, and care for people with mental illnesses.

Compulsory outpatient treatment Allows people with mental illness to be treated in the community without their consent.

Consumer choice Movement for self-determination and control over the type of treatment to be provided to persons with mental illness.

Co-occurring disorders Having another disorder such as alcoholism or drug addiction along with mental illness; also known as dual diagnosis.

Custodial care Minimal care, providing food and shelter but no active treatment.

Dangerousness The condition of posing a significant risk of hurting oneself or another; used as a standard in determining whether involuntary treatment is justified.

Deinstitutionalization The movement to shift psychiatric patients from psychiatric hospitals into the community.

DHHS Department of Health and Human Services in the federal government.

Disease model of mental illness Places a focus on finding and treating the causes of emotional, behavioral, and/or organic dysfunction, with an approach based on diagnosis, treatment, and cure.

DSM IV The *Diagnostic and Statistical Manual of Mental Disorders* is published by the APA and provides a guide for diagnosis of mental illness. It has been revised several times.

ECT Electric shock treatment uses electrodes to introduce a shock to the brain; it is primarily used to treat people with major depression and remains controversial.

Eugenics movement Movement of the late nineteenth and early twentieth century that held that insanity could be inherited.

HMO A health maintenance organization; a form of managed care that is paid for on a *capitated,* or per person, basis. Mental health care may be provided through an HMO arrangement.

ICD 10 *Manual of the International Statistical Classification of Disease, Injuries, and Causes of Death;* serves as a tool for diagnosis.

Inpatient care Care given in an inpatient setting, such as a hospital, where patients stay overnight to receive treatment.

Involuntary treatment Forced treatment; the patient does not have the right to refuse treatment. Generally requires a standard of dangerousness to self or others.

IPP Individual Program Plan As part of the movement to provide better care and treatment, the provision of a plan of treatment written for the individual.

Joint Commission on Mental Illness and Health Established under the 1955 Mental Health Study Act, the commission conducted a survey and recommended the establishment of community mental health centers.

Lanterman-Petris-Short Act Legislation passed in California in 1968; considered revolutionary for setting up strict criteria and legal procedures for involuntary hospitalization.

Lobotomy A surgical procedure, common from the 1930s to 1950s, that produced damage to the frontal lobe of the brain. It was believed that this procedure had a curative effect on people with mental illness, but there were many side effects, and the procedure was discredited in the 1950s.

Long-term care Services provided through an extended care plan. These services may include such things as adult day services.

Managed care A system to provide health or mental health care by connecting health insurance with the delivery of care.

Medicaid The shared federal and state health insurance program begun in 1965 that provides health and mental health care to low-income people.

Medicare The federal health insurance program begun in 1965 that provides health care to the elderly and some people with disabilities.

Mental disabilities A wide array of mental illnesses and disabilities, including severe mental illnesses such as schizophrenia, bipolar disorder, clinical depression, and phobias.

Mental health policy All governmental activities specifically concerned with the prevention, treatment, and living situations of people with mental disorders.

Mental health programs Programs that provide mental health services to people with mental health problems.

Moral treatment A form of treatment established in the late eighteenth century that was based on providing a supportive and sympathetic environment to assist healing.

NAMI National Alliance for the Mentally Ill; an advocacy organization that was founded in 1979 by parents of persons with mental illness.

National Institute of Mental Health A federal agency within the National Institutes of Health that provides research funding for mental health.

National Mental Health Act Passed in 1946, the act brought the federal government into mental health policy in a significant way and created new federal grants.

Outpatient care Care given in an outpatient setting; patients do not need to stay overnight in order to receive treatment.

Parity Laws that require private health insurance plans to provide equal coverage for physical health and selected mental health conditions, including serious mental disabilities in adults and serious emotional disturbances in children.

President's Commission on Mental Health An advisory body that makes recommendations concerning mental health to the president. Many recommendations have been incorporated into legislation.

Primary-care physician The first level of physician care; usually a general practitioner, family physician, or pediatrician.

Psychiatrist A medical doctor who specializes in treating mental disorders.

Psychotropic drugs Drugs that affect brain chemistry and are used to treat mental illness.

Public policy analysis Systematic investigation of public action.

Realignment In 1991, a major reform of health services in California, including programmatic, governance, and fiscal changes; it took mental health funding out of the general fund and provided a stable funding source through the sales tax and vehicle license fees.

Recovery A major element in the arena of mental health reform. Emphasizing the process of recovering from a mental illness with or without professional help.

Resilience Strengths that may promote health and healing. Involves the interaction of biological, psychological, and environmental processes.

SAMHSA Substance Abuse and Mental Health Services Administration; a federal agency within the Department of Health and Human Services that provides guidance and funding for mental health and substance abuse programs.

SMI Serious mental illness; includes any psychiatric disorder present during the past year that seriously interfered with one or more aspects of a person's daily life.

Stigmatization Being perceived as significantly different in a negative way.

Suicide Taking one's own life. A major public health problem in the United States and many other countries.

Index

About the Author

Donna R. Kemp is professor of political science and public administration at California State University, Chico. She has published numerous articles on public personnel and on health and mental health policy and services. She is author or editor of five books on topics related to mental health policy and personnel, and the recipient of three Fulbright Fellowships: to New Zealand in 1989–1990, Lithuania in 1997, and Latvia in 2004.